TECHNICAL ENGINEERING AND DESIGN GUIDES AS ADAPTED FROM THE US ARMY CORPS OF ENGINEERS, NO. 9

SETTLEMENT ANALYSIS

Published by
ASCE Press
American Society of Civil Engineers
345 East 47th Street
New York, New York 10017-2398

ABSTRACT

The U.S. Army Corps of Engineers Manual, Settlement Analysis, presents guidelines for calculation of vertical displacements and settlement of soil under shallow foundations. As soil is a nonhomogenous porous material comprising solids, fluid, and air, this manual describes vertical displacement and settlement caused by changes in stress and water content. The manual also provides guidance for the following: tests and analyses to estimate secondary compression settlement; estimation of settlement for dynamic loads; calculation of soil movements in expansive soils; calculation of settlement in collapsible soil.

Library of Congress Cataloging-in-Publication Data

Settlement analysis.
        p.cm.—(Technical engineering and design guides as adapted from the US Army Corps of Engineers; no. 9)
    Includes index.
    ISBN 0-7844-0021-0
    1. Settlement of structures. 2. Soil-structure interaction. 3. Foundations. I. American Society of Civil Engineers. II. United States. Army. Corps of Engineers. III. Series. IV. Series: Technical engineering and design guides as adapted from the U.S. Army Corps of Engineers; no. 9.
TA775.S4253 1994   94-37791
624.1—dc20   CIP

TABLE OF CONTENTS

clastic

Chapter 5. Applications with Unstable Foundation Soil

Chapter 6. Coping with Soil Movements

Appendices

Appendix C Stress Distribution in Soil

Appendix D Elastic Parameters

Appendix E Laboratory Consolidometer Tests

Appendix F Computer Program VDISPL

Appendix G Notation

Index

DEPARTMENT OF THE ARMY
U.S. Army Corps of Engineers
WASHINGTON, D.C. 20314-1000

REPLY TO
ATTENTION OF:

2 5 MAY 1994

Mr. James W. Poirot
President, American Society
  of Civil Engineers
345 East 47th Street
New York, New York 10017

Dear Mr. Poirot:

I am pleased to furnish the American Society of Civil
Engineers (ASCE) a copy of the U. S. Army Corps of Engineers
Engineering Manual, EM 1110-2-1904, Settlement Analysis. The
Corps uses this manual to provide information on calculating
vertical displacements and settlement of soil under embankments
and shallow foundations.

I understand that ASCE plans to publish this manual for
public distribution. I believe this will benefit the civil
engineering community by improving transfer of technology between
the Corps and other engineering professionals.

Sincerely,

Arthur E. Williams
Lieutenant General, U. S. Army
Commanding

CHAPTER 1

INTRODUCTION

1-1. Purpose and Scope

This manual presents guidelines for calculation of vertical displacements and settlement of soil under shallow foundations (mats and footings) supporting various types of structures and under embankments.

A. CAUSES OF SOIL DISPLACEMENT. Soil is a nonhomogeneous porous material consisting of three phases: solids, fluid (normally water), and air. Soil deformation may occur by change in stress, water content, soil mass, or temperature. Vertical displacement and settlement caused by change in stress and water content are described in this manual. Limitations of these movements required for different structures are described in Chapter 2.

(1) Elastic Deformation. Elastic or immediate deformation caused by static loads is usually small, and it occurs essentially at the same time these loads are applied to the soil. Guidance for tests and analyses to estimate immediate settlements of foundations, embankments, pavements, and other structures on cohesionless and cohesive soils for static loading conditions is given in Sections I and II of Chapter 3.

(2) Consolidation. Time-delayed consolidation is the reduction in volume associated with a reduction in water content, and it occurs in all soils. Consolidation occurs quickly in coarse-grained soils such as sands and gravels, and it is usually not distinguishable from elastic deformation. Consolidation in fine-grained soils such as clays and organic materials can be significant and take considerable time to complete. Guidance for tests and analyses to estimate consolidation settlement of foundations, embankments, pavements, and other structures on cohesive soil for static loading conditions is given in Section III of Chapter 3.

(3) Secondary Compression and Creep. Secondary compression and creep are associated with the compression and distortion at constant water content of compressible soils such as clays, silts, organic materials, and peat. Guidance for tests and analyses to estimate secondary compression settlement is given in Section IV of Chapter 3.

(4) Dynamic Forces. Dynamic loads cause settlement from rearrangement of particles, particularly in cohesionless soil, into more compact positions. Guidance to estimate settlement for some dynamic loads is given in Chapter 4.

(5) Expansive Soil. Expansive soil contains colloidal clay minerals, such as montmorillonite, that experience heave and shrinkage with changes in the soil water content. Guidance for calculation of soil movements in expansive soil is given in Section I of Chapter 5.

(6) Collapsible Soil. Collapsible soil usually consists of cohesive silty sands with a loose structure or large void ratio. The cohesion is usually caused by the chemical bonding of particles with soluble compounds such as calcareous or ferrous salts. Collapse occurs when the bonds between particles are dissolved. Guidance for calculation of settlement in collapsible soil is given in Section II of Chapter 5.

B. COPING WITH SOIL MOVEMENTS. Soil movements may be minimized by treating the soil prior to construction by numerous methods such as removing poor soil and replacing it with suitable soil, precompression of soft soil, dynamic consolidation of cohesionless soil, and chemical stabilization or wetting of expansive or collapsible soil. Foundations may be designed to tolerate some differential movements. Remedial techniques such as underpinning with piles, grouting, and slabjacking are available to stabilize and repair damaged foundations. Methods for minimizing and coping with settlement are given in Chapter 6.

C. LIMITATIONS OF THE MANUAL. This manual excludes settlement caused by subsidence and undermining by tunnels, subsidence due to buried karst features or cavities, thermal effects of structures on permafrost, effects of frost heave, loss in mass from erosion, loss of ground from rebound and lateral movement in adjacent excavations, and loss of support caused by lateral soil movement from landslides, downhill creep, and shifting retaining walls.

(1) Horizontal Deformation. Horizontal deformation of structures associated with vertical deformations may also occur, but such analysis is complex and beyond the scope of this manual.

(2) Deep Foundations. Deep foundations are driven piles and drilled shafts used to transmit foundation loads to deeper strata capable of supporting the applied loads. Guidelines on settlement analysis of deep foundations is given in TM 5-809-7, "Design of Deep Foundations (Except Hydraulic Structures)."

(3) Landfills. Settlement of domestic and hazardous landfills are unpredictable and cannot be readily estimated using techniques presented in this manual.

1-2. Applicability

This manual applies to all Corps of Engineers field operating activities. Applications include, but are not limited to, design analysis of alternatives for new construction, analyses for rationalizing in-service performance, forensic investigations, and damage assessments and repair/rehabilitation design.

1-3. References

Standard references pertaining to this manual are listed in Appendix A, References. Each reference is identified in the text by the designated Government publication number or performing agency. Additional reading materials are listed in the Bibliography and are indicated throughout the manual by numbers (e.g., item 1, 2) that correspond to similarly numbered items in Appendix B.

1-4. Recision

This manual supersedes EM 1110-2-1904, "Settlement Analysis", Chapter 4, dated January 1953.

1-5. General Considerations and Definitions

Placement of an embankment load or structure on the surface of a soil mass introduces stress in the soil that causes the soil to deform and leads to settlement of the structure. It is frequently necessary to estimate the differential and total vertical soil deformation caused by the applied loads. Differential movement affects the structural integrity and performance of the structure. Total deformation is significant relative to connections of utility lines to buildings, grade and drainage from structures, minimum height specifications of dams (i.e., freeboard), and railroad and highway embankments. Soils and conditions described in Table 1-1 require special consideration to achieve satisfactory design and performance. Early recognition of these problems is essential to allow sufficient time for an adequate field investigation and preparation of an appropriate design.

A. PRECONSOLIDATION STRESS. The preconsolidation stress or maximum effective past pressure σ_p' experienced by a foundation soil is a principle factor in determining the magnitude of settlement of a structure supported by the soil. σ_p' is the maximum effective stress to which the in situ soil has been consolidated by a previous loading; it is the boundary between recompression and virgin consolidation, which are described in Section III, Chapter 3. Pressures applied to the foundation soil that exceed the maximum past pressure experienced by the soil may cause substantial settlement. Structures should be designed, if practical, with loads that maintain soil pressures less than the maximum past pressure.

(1) Geological Evidence of a Preconsolidation Stress. Stresses are induced in the soil mass by past history such as surcharge loads from soil later eroded away by natural causes, lowering of the groundwater table, and desiccation by drying from the surface.

(a) Temporary groundwater levels and lakes may have existed, causing loads and overconsolidation compared with existing effective stresses.

(b) Desiccation of surface soil, particularly cyclic desiccation due to repeated wetting and drying, creates significant microscale stresses, which in turn cause significant preconsolidation effects. Such effects include low void ratios as well as fissures and fractures, high density, high strength, and high maximum past pressures measured in consolidation tests.

(c) A high preconsolidation stress may be anticipated if $N/15 \cdot 1/\sigma_{oz} > 0.25$ where N is the blowcount from standard penetration test (SPT) results and σ_{oz} (tons/square foot or tsf) is the total overburden pressure at depth z (table 3-2, TM 5-818-1).

(2) Evaluation from Maximum Past Thickness. Local geologic records and publications when available should be reviewed to estimate the maximum past thickness of geologic formations from erosion events, when and amount of material removed, glacial loads, and crustal tilt.

(a) The minimum local depth can sometimes be determined from transvalley geologic profiles if carried sufficiently into abutment areas to be beyond the influence of valley erosion effects.

(b) The maximum past pressure at a point in an in situ soil is estimated by multiplying the unit wet soil weight (approximately 0.06 tsf) by the total estimated past thickness of the overlying soil at that point.

(c) Results of the cone penetration test (CPT) may be used to evaluate the thickness of overburden soil removed by erosion if the cone tip resistance q_c increases linearly with depth (refer to Figure 7 in item 56). The line of q_c versus depth is extrapolated back above the existing surface of the soil to the elevation where q_c is zero assuming the original cohesion is

Table 1-1. Problem Soils and Conditions

Soil/Condition (1)	Description (2)
a. Problem Soils	
Organic	Colloids or fibrous materials such as peats, organic silts, and clays of many estuarine, lacustrine, or fluvial environments are generally weak and will deform excessively under load. These soils are usually not satisfactory for supporting even very light structures because of excessive settlements.
Normally consolidated clays	Additional loads imposed on soil consolidated only under the weight of the existing environment will cause significant long-term settlements, particularly in soft and organic clays. These clays can be penetrated several centimeters by the thumb. The magnitude and approximate rate of settlement should be determined by methods described in Section III, Chapter 3, in order to determine acceptability of settlements for the function and characteristics of the structure. Bottoms of excavations may heave and adjoining areas settle unless precautions are taken to prevent such movement.
Sensitive clays	The ratio of undisturbed to remolded strength is the sensitivity of a clay. Clays having remolded strengths 25 percent or less of the undisturbed strength are considered sensitive and subject to excessive settlement and possible catastrophic failure. Such clays preconsolidated by partial desiccation or erosion of overlying soil may support shear stresses caused by foundation loads if these loads are well within the shear strength of the clay. Refer to paragraph 3-12 on apparent preconsolidation for analysis of settlement.
Swelling and shrinking clays and shales	Clays, especially those containing montmorillonite or smectite, expand or contract from changes in water content and are widely distributed throughout the United States and the world. Clay shales may swell significantly following stress relief as in a cut or excavation and following exposure to air. Foundations in these soils may have excessive movements unless the foundation soil is treated or provisions are made in the design to account for these movements or swell pressures developed in the soil on contact with moisture. Refer to Section I, Chapter 5, for details on analysis of heave and shrinkage.
Collapsible soils	The open porous structure of loosely deposited soil such as silty clays and sands with particles bonded with soluble salts may collapse following saturation. These soils are often strong and stable when dry. Undisturbed samples should be taken to accurately determine the in situ density. Refer to Section II, Chapter 5, for details on settlement analysis.
Loose granular soils	All granular soils are subject to some densification from vibration, which may cause significant settlement and liquefaction of soil below the water table; however, minor vibration, pile driving, blasting, and earthquake motion in loose to very loose sands may induce significant settlement. Limits to potential settlement and applicable densification techniques should be determined. Refer to Chapter 4 for analysis of dynamic settlements in these soils.
Glacial tills	Till is usually a good foundation soil, but boulders and soft layers may cause problems if they are undetected during the field investigation.
Fills	Unspecified fills placed randomly with poor compaction control can settle significantly and provide unsuitable foundation soil. Fills should usually be engineered materials of low plasticity index < 12 and liquid limit < 35. Suitable materials of the Unified Soil Classification System include GW, GM, GC, GP, SW, SP, SM, SC, and CL soils. Compaction beneath structures to $\geq$ 92% of optimum density for cohesive fill or 95 percent for cohesionless fill using ASTM Standard Test Methods D 1557 has provided highly successful constructability and in-service performance. Refer to EM 1110-2-1911 for construction control of earth and rockfill dams.
b. Problem Conditions	
Meander loops and cutoffs	Soils that fill abandoned waterways are usually weak and highly compressible. The depth of these soils should be determined and estimates made of potential settlement early in design to allow time for development of suitable measures for treating the soil or accommodating settlement.
Landslides	Potential landslides are not easily detected, but evidence of displacement such as bowed trees and tilted or warped strata should be noted. Sensitive clays and cutting action of eroding rivers significantly increase the risk of landslides. Slopes and excavations should be minimized,

Table 1-1. Continued

Soil/Condition (1)	Description (2)
	b. Problem Conditions
	seasonal variations in the local water table considered in the design, and suitable arrangements for drainage provided at the top and toe of slopes.
Kettle holes	The retreating continental ice sheet left large blocks of ice that melted and left depressions, which eventually filled with peat or soft organic soils. Lateral dimensions can vary from a few to several hundred feet. Depths of kettle holes usually do not exceed 40 percent of lateral dimensions and can sometimes be identified as shallow surface depressions.
Mined areas and sinkholes	Voids beneath the surface soil may lead to severe ground movements and differential settlement from subsidence or caving. Sinkholes are deep depressions formed by the collapse of the roofs of underground caverns such as in limestone. Maps of previous mined areas are helpful when available. Published geological data, nondestructive in situ tests, and past experience help indicate the existence of subsurface cavities. Investigations should be thorough to accurately determine the existence and location of any subsurface voids.
Lateral soil distortions	Lateral distortions are usually not significant, but they can occur in highly plastic soils near the edge of surface loads. These distortions can adversely affect the performance of foundations of structures and embankments. Driven piles can cause large lateral displacements and excessive pressures on retaining walls.
Downdrag	Compression of fills or consolidation of soft soil adjacent to wall footings or piles causes downdrag on the footing or pile. This leads to substantial loads at the base of the foundation that can exceed the bearing capacity of the underlying soil supporting the footing or pile. Failure of the foundation can occur with gross distortion.
Vibrations	Cohesionless soil, especially loose sands and gravels, can densify and settle when subject to machine vibration, blasts, and earthquakes. Distortion with negligible volume change can occur in loose, saturated sands due to liquefaction. Low-level sustained vibration can densify saturated sands.

Note. Based on information from Canadian Geotechnical Society (1985).

zero. The difference in elevation from where q_c is zero and the existing elevation is the depth of overburden removed by erosion. This depth times the unit wet weight γ is the total maximum past pressure σ_p. The cohesion for many clays is not zero but contributes to a q_c approaching 1 tsf. Extrapolating the line above the existing ground surface to $q_c = 1$ tsf produces a more conservative depth of overburden clay soil. This latter estimate of overburden depth is recommended.

(3) Evaluation from Overconsolidation Ratio. The preconsolidation stress σ'_p may be evaluated from the overconsolidation ratio (OCR), σ'_p/σ'_{oz}, where σ'_{oz} is the effective vertical overburden pressure at depth z.

(a) The initial vertical effective pressure in a saturated soil mass before placement of an applied load from a structure is given by

$$\sigma'_{oz} = \gamma z - u_w \qquad (1\text{-}1)$$

where σ'_{oz} = initial vertical effective stress at depth z, in tsf; γ = saturated unit weight of soil mass at depth z, in tsf; z = depth, in ft; and u_w = pore water pressure, in tsf. u_w usually is the hydrostatic pressure $\gamma_w \cdot z_w$

where γ_w is the unit weight of water, 0.031 tsf, and z_w is the height of a column of water above depth z. γz is the total overburden pressure σ_{oz}.

(b) The overconsolidation ratio has been related empirically with the coefficient of earth pressure at rest K_o, $\sigma'_{hz}/\sigma'_{oz}$, and the plasticity index PI in figure 3-21, TM 5-818-1. σ'_{hz} is the effective horizontal pressure at rest at depth z. Normally consolidated soil is defined as soil with OCR = 1. Overconsolidated soil is defined as soil with OCR > 1.

(c) The results of pressuremeter tests (PMT) may be used to evaluate the effective horizontal earth pressure σ'_{hz}. K_o may be evaluated if the effective vertical overburden pressure σ'_{oz} at depth z is known and the OCR is estimated as above.

(4) Laboratory Tests. The preconsolidation stress may be calculated from results of consolidation tests on undisturbed soil specimens, see paragraph 3-12.

(a) A high preconsolidation stress may be anticipated if the natural water content is near the plastic limit PL or below or if $C_u/\sigma_{oz} > 0.3$ where C_u is the undrained shear strength (table 3-2, TM 5-818-1).

(b) An empirical relationship between the preconsolidation stress and liquidity index as a function of clay sensitivity, ratio of undisturbed to re-molded undrained shear strength, is given in Figure 1-1. The preconsolidation stress may also be estimated from (NAVFAC DM-7.1)

$$\sigma_p' = \frac{C_u}{0.11 + 0.0037 \text{ PI}}$$ (1-2)

where σ_{oz}' = preconsolidation stress, in tsf; C_u = undrained shear strength, in tsf; and PI = plasticity index, in percent.

B. PRESSURE BULB OF STRESSED SOIL.
The pressure bulb is a common term that represents the volume of soil or zone below a foundation within which the foundation load induces appreciable stress. The stress level at a particular point of soil beneath a foundation may be estimated by the theory of elasticity.

(1) Applicability of the Theory of Elasticity.
Earth masses and foundation boundary conditions correspond approximately with the theory of plasticity (item 52).

(2) Stress Distribution.
Various laboratory, prototype, and full-scale field tests of pressure cell measurements in response to applied surface loads on homogeneous soil show that the measured soil vertical stress distribution corresponds reasonably well to analytical models predicted by linear elastic analysis for similar boundary conditions.

(a) The Boussinesq method is commonly used to estimate the stress distribution in soil. This distribution indicates that the stressed zone decreases toward the edge of the foundation and becomes negligible (less than 10 percent of the stress intensity) at depths of about 6 times the width of an infinite strip or 2 times the width of a square foundation; see Figure 1-2.

(b) The recommended depth of analysis is at least twice the least width of the footing or mat foundation, 4 times the width of infinite strips or embankments, or the depth of incompressible soil, whichever comes first.

(c) The distribution of vertical stress in material overlain by a much stiffer layer is more nearly determined by considering the entire mass homogeneous rather than a layered elastic system.

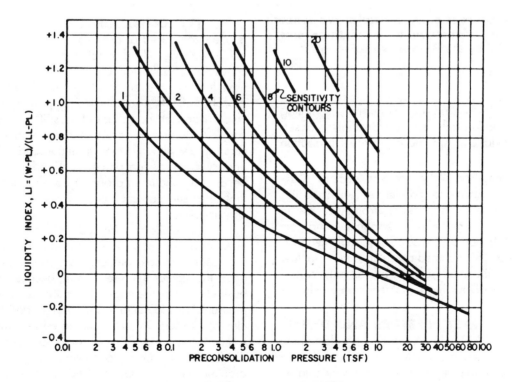

W = WATER CONTENT, PERCENT
PL = PLASTIC LIMIT, PERCENT
LL = LIQUID LIMIT, PERCENT

Figure 1-1. Preconsolidation stress as a function of liquidity index LI and clay sensitivity (ratio of undisturbed to remolded shear strength) (After NAVFAC DM 7.1)

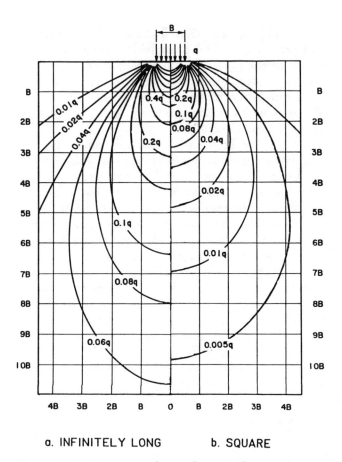

a. INFINITELY LONG   b. SQUARE

Figure 1-2. Contours of equal vertical stress beneath a foundation in a semi-infinite elastic solid by the Boussinesq solution

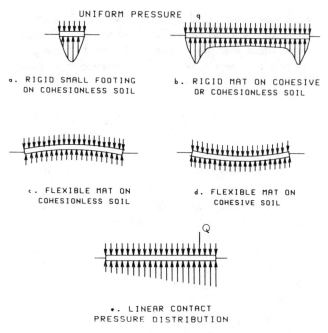

Figure 1-3. Relative distribution of soil contact pressures and displacements of rigid and flexible mats or footings on cohesionless and cohesive soils

(*d*) Methods and equations for estimating stresses in foundation soils required for analysis of settlement are provided in Appendix C, Stress Distribution in Soil.

(3) Applicability to Settlement Calculations. The ability to predict settlements using elastic theory depends much more strongly on the in situ nonlinearity and material inhomogeneity than errors in the distribution of stresses. These settlements directly depend on the assumed constitutive material law and on the magnitude of the required soil parameters. Refer to Appendix D for further information on elasticity theory.

C. CONTACT PRESSURE AND DEFORMATION PATTERN. The shape of the deformation pattern varies, depending on the flexibility of the foundation and type of soil. Figure 1-3 illustrates the relative distribution of soil contact pressures and displacements on cohesionless and cohesive soil. Linear contact pressure distributions from uniformly applied pressure q are often assumed for settlement analysis (Figures 1-3c and 1-3d). An applied load Q may cause an unequal linear soil contact pressure distribution (Figure 1-3e).

(1) Cohesionless Soil. Cohesionless soil is often composed of granular or coarse-grained materials with visually detectable particle sizes and little cohesion or adhesion between particles. These soils have little or no strength when unconfined and little or no cohesion when submerged. Apparent adhesion between particles in cohesionless soil may occur from capillary tension in pore water. Settlement usually occurs rapidly with little long-term consolidation and secondary compression or creep. Time rate effects may become significant in proportion to the silt content such that the silt content may dominate consolidation characteristics.

(*a*) Uniformly loaded rigid foundations (footings of limited size or footings on cohesionless soil) may cause less soil contact pressure near the edge than near the center (Figure 1-3a), because this soil is pushed aside at the edges due to the reduced confining pressure. This leads to lower strength and lower modulus of elasticity in soil near the edge compared with soil near the center. The parabolic soil contact pressure distribution may be replaced with a saddle-shaped distribution (Figure 1-3b) for rigid footings or mats if the soil pressure does not approach the allowable bearing capacity.

(*b*) The distortion of a uniformly loaded flexible footing, mat, or embankment on cohesionless soil will be concave downward (Figure 1-3c) because the soil near the center is stressed under higher confining pressure such that the modulus of elasticity of the soil is higher than near the edge.

(c) The theory of elasticity is not applicable to cohesionless soil when the stress or loading increment varies significantly throughout the soil such that an equivalent elastic modulus cannot be assigned. Semi-empirical and numerical techniques have been useful in determining equivalent elastic parameters at points in the soil mass based on stress levels that occur in the soil.

(2) Cohesive Soil. Cohesive soil often contains fine-grained materials consisting of silts, clays, and organic material. These soils have significant strength when unconfined and air-dried. Most cohesive soil is relatively impermeable, and when loaded it deforms in a manner similar to gelatin or rubber, i.e., the undrained state. Cohesive soils may include granular materials with bonding agents between particles such as soluble salts or clay aggregates. Wetting of soluble agents bonding granular particles may cause settlement in loose or high-void-ratio soil. Refer to Section II, Chapter 5, for evaluation of settlement in collapsible soil.

(a) A uniform pressure applied to a rigid foundation on cohesive soil (Figure 1-3b), can cause the soil contact pressure to be maximum at the edge and decrease toward the center because additional contact pressure is generated to provide stress that shears the soil around the perimeter.

(b) A uniform pressure applied to a flexible foundation on cohesive soil (Figure 1-3d) causes greater settlement near the center than near the edge because the cumulative stresses are greater near the center as a result of the pressure bulb stress distribution indicated in Figure 1-2. Earth pressure measurements from load cells beneath a stiffening beam supporting a large but flexible ribbed mat also indicated large-perimeter earth pressures resembling a saddle-shaped pressure distribution similar to Figure 1-3b (item 29).

(c) Elastic theory has been found to be useful for evaluation of immediate settlement when cohesive soil is subjected to moderate stress increments. The modulus of elasticity is a function of the soil shear strength and often increases with increasing depth in proportion with the increase in soil shear strength.

(d) Cohesive soil subject to stresses exceeding the maximum past pressure of the soil may settle substantially from primary consolidation and secondary compression and creep.

D. SOURCES OF STRESS. Sources of stress in soil occur from soil weight, surface loads, and environmental factors such as desiccation from drought, wetting from rainfall, and changes in depth to groundwater.

(1) Soil Weight. Soil strata with different unit weights alter the stress distribution. Any change in

Table 1-2. Some Typical Loads on Building Foundations

Structure (1)	Line load (tons/ft) (2)	Column load (tons) (3)
Apartments	0.5 to 1	30
Individual housing	0.5 to 1	<5
Warehouses	1 to 2	50
Retail spaces	1 to 2	40
Two-story buildings	1 to 2	40
Multistory buildings	2 to 5	100
Schools	1 to 3	50
Administration buildings	1 to 3	50
Industrial facilities	—	50

total stress results in changes in effective stress and pore pressure. In a saturated soil, any sudden increase in applied total stress results in a corresponding pore pressure increase (Equation 1-1). This increase may cause a flow of water out of the soil deposit, a decrease in pore pressure, and an increase in effective stress. Changes in pore water pressure such as the raising or lowering of water tables also lead to a reduction or increase in effective stress.

(2) Surface Loads. Loads applied to the surface of the soil mass increase the stress within the mass. The pressure bulb concept (Figure 1-2) illustrates the change in vertical stress within the soil mass. Placement of a uniform pressure over a foundation with a minimum width much greater than the depth of the soil layer will cause an increase of vertical stress in the soil approximately equal to the applied pressure.

(3) Rules of Thumb for Static Loads. Preliminary settlement analyses are sometimes performed before the structural engineer and architect are able to furnish the design load conditions.

(a) Some rules of thumb for line and column loads for buildings described in Table 1-2 are based on a survey of some engineering firms. Tall multistory structures may have column loads exceeding 1000 tons. Column spacings are often 20 to 25 ft or more. The average pressure applied per story of a building often varies from 0.1 to 0.2 tsf. Refer to TM 5-809-1/AFM 88-3, chapter 1, "Load Assumptions for Buildings," for estimating unfactored structural loads.

(b) Vertical pressures from embankments may be estimated from the unit wet weight times height of the fill.

(c) Vertical pressures from locks, dams, and retaining walls may be estimated by dividing the structure into vertical sections of constant height and evaluating the unit weight times the height of each section.

CHAPTER 2

LIMITATIONS OF SETTLEMENT

2-1. General

Significant aspects of settlement from static and dynamic loads are total and differential settlement. Total settlement is the magnitude of downward movement. Differential settlement is the difference in vertical movement between various locations of the structure and it distorts the structure. Conditions that cause settlement were described in Table 1-1. Limitations to total and differential settlement depend on the function and type of structure.

2-2. Total Settlement

Many structures can tolerate substantial downward movement or settlement without cracking (Table 2-1); however, total settlement should not exceed 2 in. for most facilities. A typical specification of total settlement for commercial buildings is 1 in. (item 35). Structures such as solid reinforced concrete foundations supporting smokestacks, silos, and towers can tolerate larger settlements up to 1 ft.

Some limitations of total settlement are as follows:

A. UTILITIES. Total settlement of permanent facilities can harm or sever connections to outside utilities such as water, natural gas, and sewer lines. Water and sewer lines may leak, contributing to localized wetting of the soil profile and aggravating differential displacement. Leaking gas from breaks caused by settlement can lead to explosions.

B. DRAINAGE. Total settlement reduces or interferes with drainage of surface water from permanent facilities, contributes to wetting of the soil profile with additional differential movement, and may cause the facility to become temporarily inaccessible.

C. SERVICEABILITY. Relative movement between the facility and surrounding soil may interfere with serviceability of entry ways.

D. FREEBOARD. Total settlement of embankments, levees, and dams reduces freeboard and volume of water that may be retained. The potential for flooding is greater during periods of heavy rainfall. Such settlement also alters the grade of culverts placed under roadway embankments.

2-3. Differential Settlement

Differential settlement, which causes distortion and damages in structures, is a function of the uniformity of the soil, stiffness of the structure, stiffness of the soil, and distribution of loads within the structure. Limitations to differential settlement depend on the application. Differential settlements should not usually exceed 1/2 in. in buildings, or cracking and structural damage may occur. Differential movements between monoliths of dams should not usually exceed 2 in., or leakage may become a problem. Embankments, dams, one- or two-story facilities, and multistory structures with flexible framing systems are sufficiently flexible that their stiffness often need not be considered in settlement analysis. Pavements may be assumed to be completely flexible.

A. TYPES OF DAMAGE. Differential settlement may lead to tilting that can interfere with adjacent structures and disrupt the performance of machinery and people. Differential settlement can cause cracking in the structure, distorted and jammed doors and windows, uneven floors and stairways, and other damage to houses and buildings. Differential movement may lead to misalignment of monoliths and reduce the efficiency of waterstops. Refer to chapter 2, EM 1110-2-2102, for guidance on selection of waterstops. Widespread cracking can impair the structural integrity and lead to collapse of the structure, particularly during earthquakes. The height of a wall for a building that can be constructed on a beam or foundation without cracking is related to the deflection/span length Δ/L ratio and the angular distortion β described later.

B. DEFLECTION RATIO. The deflection ratio Δ/L is a measure of the maximum differential movement Δ in the span length L (Figure 2-1). The span length may be between two adjacent columns, L_{SAG} or L_{HOG} (Figure 2-1a).

(1) Table 2-2 provides limiting observed deflection ratios for some buildings.

(2) Design Δ/L ratios are often greater than 1/600, but the stiffness contributed by the components of an assembled brick structure, for example, helps maintain actual differential displacement/span length

Table 2-1. Maximum Allowable Average Settlement of Some Structures (Data from Item 53)

Type of structure (1)	Settlement (in.) (2)
Plain brick walls (length/height ≥ 2.5)	3
Plain brick walls (length/height ≤ 1.5	4
Framed structure	4
Reinforced brick walls and brick walls with reinforced concrete	6
Solid reinforced concrete foundations supporting smokestacks, silos, towers, etc.	12

Table 2-2. Some Limiting Deflection Ratios (After Items 17, 53, 65)

Structure (1)	Deflection Ratio (Δ/L)	
	Sand and hard clay (2)	Plastic clay (3)
Buildings with plain brick walls (length/height ≥ 3)	1/3333	1/2500
Buildings with plain brick walls (length/height ≥ 5)	1/2000	1/1500
One story mills; between columns for brick-clad column frames	1/1000	1/1000
Steel and concrete frame	1/500	1/500

ratios near those required for brick buildings, (Table 2-2) to avoid cracking.

   (3) Circular steel tanks can tolerate Δ/L ratios greater than 1/200 depending on the settlement shape (item 13).

   C. ANGULAR   DISTORTION. Angular distortion $\beta = \delta/l$ is a measure of differential movement δ between two adjacent points separated by the distance l (Figure 2-1).

   (1) Initiation of Damage. Table 2-3 shows limits to angular distortion for various types of

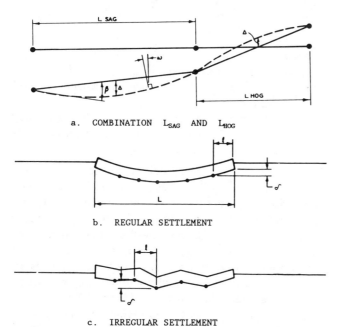

a. COMBINATION L_{SAG} AND L_{HOG}

b. REGULAR SETTLEMENT

c. IRREGULAR SETTLEMENT

Figure 2-1. Angular distortion $\beta = \delta/l$ and deflection ratio Δ/L for settling (sagging) and heaving (hogging) profiles

structures without cracking based on field surveys of damage.

   (*a*) A safe limit for no cracking in properly designed and constructed steel or reinforced concrete frame structures is angular distortion $\beta = 1/500$. Cracking should be anticipated when β exceeds 1/300. Considerable cracking in panels and brick walls and structural damage is expected when β is less than 1/150.

   (*b*) Tilting can be observed if $\omega > 1/250$; it must be limited to clear adjacent buildings, particularly in high winds. The angle of tilt is indicated by ω (Figure 2-1a).

   (*c*) Slower rates of settlement increase the ability of structures to resist cracking.

   (*d*) Unreinforced concrete masonry unit construction is notably brittle and cracks at relatively low angular distortion values as shown in Table 2-3. Such structures must be properly detailed and constructed to provide acceptable service at sites with even moderate differential movement potential. Consideration should be given to using a less crack-susceptible material at expansive soil sites and any other site having a significant differential movement potential.

   (2) Influence of Architecture. Facades, siding, and other architectural finishes are usually placed after a portion of the settlement has occurred. Most settlement, for example, may have already occurred for facilities on cohesionless soil, whereas, very little settlement may have occurred for facilities on compressible cohesive soil when the facade is to be placed.

   (*a*) Larger angular distortions than those shown in Table 2-3 can be accommodated if some of the settlement has occurred before installation of architectural finishes.

Table 2-3. Limiting Angular Distortions to Avoid Potential Damages
(Data from Items 53, 65, TM 5-818-1)

Situation (1)	Length / Height (2)	Allowable angular distortion ($\beta = \delta/l$) (3)
Hogging of unreinforced load-bearing walls	—	1/2000
Load-bearing brick, tile, or concrete block walls	≥ 5	1/1250
Load-bearing brick, tile, or concrete block walls	≤ 3	1/2500
Sagging of unreinforced load-bearing walls	—	1/1000
Machinery sensitive to settlement	—	1/750
Frames with diagonals	—	1/600
No cracking in buildings; tilt of bridge abutments; tall slender structures such as stacks, silos, and water tanks on a rigid mat	—	1/500
Steel or reinforced concrete frame with brick, block, plaster, or stucco finish	≥ 5	1/500
Steel or reinforced concrete frame with brick, block, plaster, or stucco finish	≤ 3	1/1000
Circular steel tanks on flexible base with floating top; steel or reinforced concrete frames with insensitive finish such as dry wall, glass, or panels	—	1/300 to 1/500
Cracking in panel walls; problems with overhead cranes	—	1/300
Tilting of high rigid buildings	—	1/250
Structural damage in buildings; flexible brick walls with length/height ratio > 4	—	1/150
Circular steel tanks on flexible base with fixed top; steel framing with flexible siding	—	1/125

(*b*) The allowable angular distortion of the structure (Table 2-3) should be greater than the estimated maximum angular distortion of the foundation (Table 2-4) to avoid distress in the structure.

D. ESTIMATION OF THE MAXIMUM ANGULAR DISTORTION. The maximum angular distortion for uniformly loaded structures on laterally uniform cohesive soil profiles occurs at the corner (Figure 2-1b). The maximum angular distortion may be estimated from the lateral distribution of calculated settlement. The maximum angular distortion for structures on sand, compacted fill, and stiff clay often occurs any-where on the foundation because the settlement profile is usually erratic (Figure 2-1c).

(1) The maximum angular distortion at a corner of a foundation shaped in a circular arc on a uniformly loaded cohesive soil for the Boussinesq stress distribution, Appendix C, is approximately

$$\beta_{max} - \frac{3 \cdot \rho}{(N_{col} - 1) \cdot l} \qquad (2\text{-}1)$$

where ρ_{max} = maximum settlement in center of mat in ft; N_{col} = number of columns in a diagonal line on the foundation; and l = distance between adjacent columns on the diagonal line in ft.

The maximum settlement may be calculated from loads on soil beneath the center of the foundation using the methodology of Chapter 3.

(2) When the potential for soil heave and nonuniform soil wetting exists, the maximum angular distortion may be the sum of the maximum settlement ρ_{max} without soil wetting and maximum potential heave S_{max} of wetted soil divided by the minimum distance between ρ_{max} and S_{max}. S_{max} may occur beneath the most lightly loaded part of the foundation such as the midpoint between diagonal columns. ρ_{max} may occur beneath the most heavily loaded part of the structure. ρ_{max} will normally only be the immediate elastic settlement ρ_i; consolidation is not expected in a soil with potential for heave in situ. Nonuniform soil wetting may be caused by leaking water, sewer, and drain lines.

Table 2-4. Empirical Correlations between Maximum Distortion Δ and Angular Distortion β (From Table 5-3, TM 5-818-1)

Soil (1)	Foundation (2)	Approximate β for $\Delta = 1$ in.[a]
Sand	Mats	1/750
Sand	Spread footings	1/600
Varved silt	Rectangular mats	1/1000 to 1/2000
Varved silt	Square mats	1/2000 to 1/3000
Varved silt	Spread footings	1/600
Clay	Mats	1/1250
Clay	Spread footings	1/1000

[a] β increases roughly in proportion with Δ. For $\Delta = 2$ in., β is about twice as large as those shown; for $\Delta = 3$ in., three times as large, etc.

(3) When the potential for soil heave and uniform wetting occurs, the maximum angular distortion will be the difference between the maximum and minimum soil heave divided by the minimum distance between these locations. The maximum and minimum heave may occur beneath the most lightly and heavily loaded parts of the structure, respectively. Uniform wetting may occur following construction of the foundation through elimination of evapotranspiration from the ground surface.

(4) When the potential for soil collapse exists on wetting of the subgrade, the maximum angular distortion will be the difference between the maximum settlement of the collapsible soil ρ_{col} and ρ_{min} divided by the distance between these points or adjacent columns. ρ_{min} may be the immediate settlement assuming collapse does not occur (no soil wetting) beneath a point. See Chapter 5 for further details on heaving and collapsing soil and Sections I and II of Chapter 3 for details on calculating immediate settlement.

E. CORRELATIONS BETWEEN DEFLECTION RATIO AND ANGULAR DISTORTION.
The deflection ratio Δ/L may be estimated from the maximum angular distortion or slope at the support by (item 65)

$$\frac{\Delta}{L} = \frac{\beta}{3} \cdot \left[\frac{1 + 3.9\,(H_w/L)^2}{1 + 2.6(H_w/L)^2} \right] \qquad (2\text{-}2)$$

where Δ = differential displacement in ft; L = span length or length between columns in ft; H_w = wall height

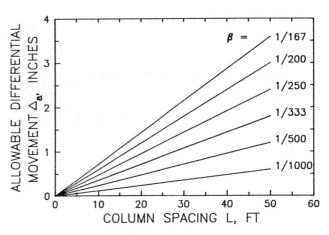

Figure 2-2. Allowable differential movement for buildings (After NAVFAC DM-7.1)

in ft; and β = angular distortion. The deflection ratio Δ/L is approximately 1/3 of the angular distortion β for short, long structures or L/H_w greater than 3.

(1) Table 2-4 illustrates empirical correlations between the maximum deflection Δ and angular distortion β for uniformly loaded mats and spread footings on homogeneous sands, silts, and clays.

(2) Figure 2-2 illustrates a relationship between the allowable differential settlement Δ_a, column spacing L, and the angular distortion β.

CHAPTER 3

EVALUATION OF SETTLEMENT FOR STATIC LOADS

3-1. General

This chapter presents the evaluation of immediate settlement in cohesionless and cohesive soils and consolidation settlement of soil for static loads. Settlement is denoted as a positive value to be consistent with standard practice.

3-2. Components of Settlement

Total settlement ρ in feet, which is the response of stress applied to the soil, may be calculated as the sum of three components

$$\rho = \rho_i + \rho_c + \rho_s \qquad (3\text{-}1)$$

where ρ_i = immediate or distortion settlement in ft; ρ_c = primary consolidation settlement in ft; and ρ_s = secondary compression settlement in ft. Primary consolidation and secondary compression settlements are usually small if the effective stress in the foundation soil applied by the structure is less than the maximum effective past pressure of the soil (paragraph 1-5a).

A. IMMEDIATE SETTLEMENT. Immediate settlement ρ_i is the change in shape or distortion of the soil caused by the applied stress.

(1) Calculation of immediate settlement in cohesionless soil is complicated by a nonlinear stiffness that depends on the state of stress. Empirical and semi-empirical methods for calculating immediate settlement in cohesionless soils are described in Section I.

(2) Immediate settlement in cohesive soil may be estimated using elastic theory, particularly for saturated clays, clay shales, and most rocks. Methods for calculating immediate settlement in cohesive soil are described in Section II.

B. PRIMARY CONSOLIDATION SETTLEMENT. Primary consolidation settlement ρ_c occurs in cohesive or compressible soil during dissipation of excess pore fluid pressure, and it is controlled by the gradual expulsion of fluid from voids in the soil leading to the associated compression of the soil skeleton. Excess pore pressure is pressure that exceeds the hydro-

static fluid pressure. The hydrostatic fluid pressure is the product of the unit weight of water and the difference in elevation between the given point and elevation of free water (phreatic surface). The pore fluid is normally water with some dissolved salts. The opposite of consolidation settlement (soil heave) may occur if the excess pore water pressure is initially negative and approaches zero following absorption and adsorption of available fluid.

(1) Primary consolidation settlement is normally insignificant in cohesionless soil and occurs rapidly because these soils have relatively large permeabilities.

(2) Primary consolidation takes substantial time in cohesive soils because they have relatively low permeabilities. Time for consolidation increases with thickness of the soil layer squared and is inversely related to the coefficient of permeability of the soil. Consolidation settlement determined from results of one-dimensional consolidation tests include some immediate settlement ρ_i. Methods for calculating primary consolidation settlement are described in Section III.

C. SECONDARY COMPRESSION SETTLEMENT. Secondary compression settlement is a form of soil creep that is largely controlled by the rate at which the skeleton of compressible soils, particularly clays, silts, and peats, can yield and compress. Secondary compression is often conveniently identified to follow primary consolidation when excess pore fluid pressure can no longer be measured; however, both processes may occur simultaneously. Methods for calculating secondary compression settlement are described in Section IV.

Section I. Immediate Settlement of Cohesionless Soil for Static Loads

3-3. Description of Methods

Settlement in cohesionless soil (see paragraph 1-5c for definition) is normally small and occurs quickly with little additional long-term compression. The six methods described herein for estimating settlement in cohesionless soil are based on data from field tests [i.e., standard penetration test (SPT), cone penetration test (CPT), dilatometer test (DMT), and pressuremeter test (PMT)]. Undisturbed samples of cohesionless soil are normally not obtainable for laboratory tests. The first four empirical and semiempirical methods—Alpan, Schultze and Sherif; modified Terzaghi and Peck; and Schmertmann approximations—were shown to provide estimates from about 1/4 to 2 times the measured settlement for 90 percent confidence based on the results of a statistical analysis (item 27). Penetration tests may not be capable of sensing effects of prestress or overconsolidation and can underestimate the stiffness that may lead to overestimated settlements (item 37).

A. ALPAN APPROXIMATION. This procedure estimates settlement from a correlation of SPT data with settlement of a 1-ft-square loading plate. The settlement of a footing of width B in feet is (item 1)

$$\rho_i = m' \cdot \left[\frac{2B}{1+B} \right]^2 \cdot \frac{\alpha_o}{12} \cdot q \qquad (3\text{-}2)$$

where ρ_i = immediate settlement in ft; m' = shape factor, $(L/B)^{0.39}$; L = length of footing in ft; B = width of footing in ft; α_o = parameter from Figure 3-1a using an adjusted blowcount N' from Figure 3-1b in in./tsf; and q = average pressure applied by footing on soil in tsf.

(1) Blowcount N. N is the average blowcount per foot in the stratum, number of blows of a 140-pound hammer falling 30 in. to drive a standard sampler (1.42-in. I.D., 2.00-in. O.D.) 1 ft. The sampler is driven 18 in. and blows are counted the last 12 in. The blowcount should be determined by ASTM Standard Test Method D 1586. Prior to 1980 the efficiency of the hammer was not well recognized as influencing the blowcount and was usually not considered in analysis.

(*a*) The measured blowcounts should be converted to 60 percent of input energy N_{60} by

$$N_{60} = N_m \cdot \frac{ER_i}{60} \qquad (3\text{-}3a)$$

$$ER_i = \frac{E_i}{E^*} \qquad (3\text{-}3b)$$

where N_{60} = blowcounts corrected to 60 percent energy ratio; N_m = blowcounts measured with available energy E_i; ER_i = measured energy ratio for the drill rig and hammer system; and E^* = theoretical SPT energy applied by a 140-pound hammer falling freely 30 in., 4200 in.-lb.

(*b*) The converted blowcount N_{60} is entered in Figure 3-1a with the calculated effective overburden

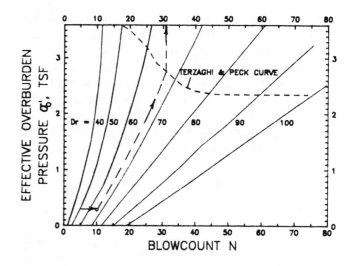

a. ADJUSTED BLOWCOUNT FROM N AND σ_o'

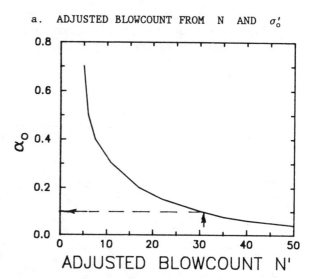

b. PARAMETER α_o FROM ADJUSTED BLOWCOUNT

Figure 3-1. Chart to apply Alpan's procedure (data from item 1)

pressure σ_o' at the base of the footing to estimate the relative density D_r. The relative density is adjusted to 100 percent using the Terzaghi-Peck curve and the adjusted blowcount N' for D_r = 100 percent. For example, if σ_o' = 0.3 tsf and N = 10, then the relative density D_r = 67 percent (Figure 3-1a). The adjusted N' is determined to be 31 for D_r = 100 percent.

(2) Parameter α_o. The adjusted blowcount is entered in Figure 3-1b to determine α_o. α_o = 0.1 in./tsf for adjusted N' = 31.

B. SCHULTZE AND SHERIF APPROXIMATION. This procedure estimates settlement from the blowcount of SPT results based on 48 field cases (item 60)

$$\rho_i = \frac{f \cdot q \cdot \sqrt{B}}{N_{ave}^{0.87} \cdot \left(1 + 0.4 \frac{D}{B}\right)} \qquad (3\text{-}4)$$

where f = influence factor from elasticity methods for isotropic half-space (Figure 3-2); H = depth of stratum below footing to a rigid base in ft; D = depth of embedment in ft; and N_{ave} = average blowcount/ft in depth H. The depth to the rigid base H should be ≤ $2B$. N_{ave} is based on measured blowcounts adjusted to N_{60} by Equations 3-3.

C. MODIFIED TERZAGHI AND PECK APPROXIMATION.
This procedure is a modification of the original Terzaghi and Peck approach to consider overburden pressure and water table (items 50 and 51)

$$\rho_i = \frac{q}{18 \cdot q_1} \qquad (3\text{-}5)$$

where q_1 = soil pressure from Figure 3-3a using corrected blowcount N' and the ratio of embedment depth D to footing width B in tsf. The corrected blowcount N'

is found from

$$N' = N \cdot Cw \cdot Cn \qquad (3\text{-}6)$$

where N = average blowcount/ft in the sand; Cw = correction for water table depth; and Cn = correction for overburden pressure (Figure 3-3b). Equation 3-5 calculates settlements 2/3 of the Terzaghi and Peck method (item 51) as recommended by Peck and Bazarra (item 50).

(1) Water Table Correction. The correction Cw is given by

$$Cw = 0.5 + 0.5 \cdot \frac{Dw}{D + B} \qquad (3\text{-}7)$$

where Dw = depth to groundwater level in ft. The correction factor $Cw = 0.5$ for a groundwater level at the ground surface. The correction factor is 1 if the sand is dry or the groundwater level exceeds the depth $D + B$ below the ground surface.

(2) Overburden Pressure Correction. The correction factor Cn is found from Figure 3-3b as a function of the effective vertical overburden pressure σ_o'.

D. SCHMERTMANN APPROXIMATION.
This procedure provides settlement compatible with field measurements in many different areas. The analysis assumes that the distribution of vertical strain is compatible with a linear elastic half-space subjected to a uniform pressure (item 55)

$$\rho_i = C_1 \cdot C_t \cdot \Delta p \cdot \sum_{i=1}^{n} \frac{\Delta z_1}{E_{si}} \cdot Izi \qquad (3\text{-}8)$$

where C_1 = correction to account for strain relief from embedment, $1 - 0.5\sigma_{od}'/\Delta p \geq 0.5$; σ_{od}' = effective vertical overburden pressure at bottom of footing or depth D in tsf; Δp = net applied footing pressure, $q - \sigma_{od}'$ in tsf; C = correction for time-dependent increase in settlement, $1 + 0.2 \cdot \log_{10}(t/0.1)$; t = time in years; E_{si} = elastic modulus of soil layer i in tsf; Δz_i = depth increment i, $0.2B$ in ft; and Izi = influence factor of soil layer i (Figure 3-4). Settlement may be calculated with the assistance of the calculation sheet (Figure 3-5). The time-dependent increase in settlement is related with creep and secondary compression as observed in clays.

(1) Influence Factor. The influence factor Iz is based on approximations of strain distributions for square or axisymmetric footings and infinitely long or plane strain footings observed in cohesionless soil, which are similar to an elastic medium such as the Boussinesq distribution (Figure 1-2). The peak value of

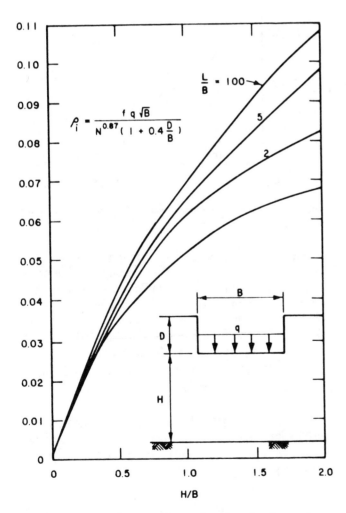

Figure 3-2. Settlement from the standard penetration test (data from item 60)

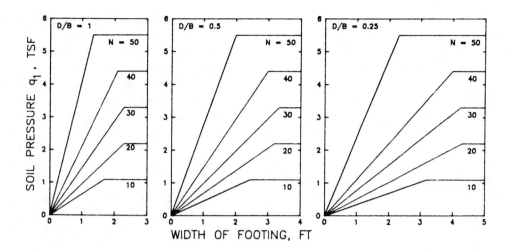

a. EVALUATION OF SOIL PRESSURE q_1 FROM CORRECTED BLOWCOUNT N'
AND EMBEDMENT DEPTH/FOOTING WIDTH RATIO D/B

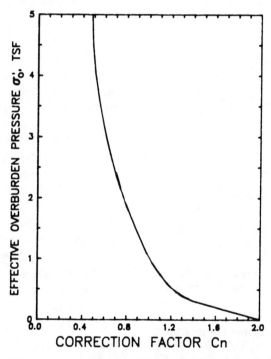

b. CORRECTION Cn FOR EFFECTIVE OVERBURDEN PRESSURE σ_o'

Figure 3-3. Charts for Modified Terzaghi and Peck Approximation. Reprinted by permission of John Wiley & Sons, Inc. from *Foundation Engineering*, 2nd Edition, Copyright © 1974 by R. B. Peck, W. E. Hanson, and T. H. Thornburn, pp 309, 312

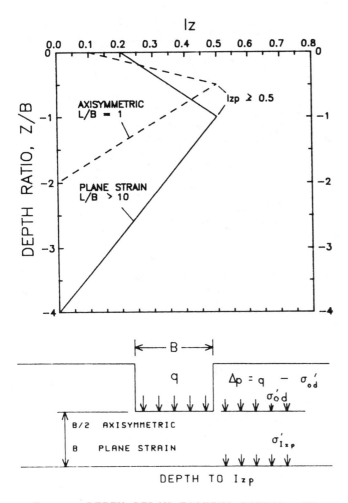

Z = DEPTH BELOW FOOTING BOTTOM, FT
B = FOOTING WIDTH, FT
Iz = DEPTH INFLUENCE FACTOR
Izp = PEAK DEPTH INFLUENCE FACTOR

Figure 3-4. Recommended strain influence factors for Schmertmann's Approximation. Reprinted with permission of the American Society of Civil Engineers from the *Journal of the Geotechnical Engineering Division*, Vol 104, 1978, "Improved Strain Influence Factor Diagram", by J. M. Schmertmann, J. P. Hartman, and P. R. Brown, p. 1134

the influence factor Izp in Figure 3-4 is (item 59)

$$Izp = 0.5 + 0.1 \left[\frac{\Delta p}{\sigma'_{Izp}} \right]^{1/2} \quad (3\text{-}9a)$$

Axisymmetric: $\quad \sigma'_{Izp} = 0.5 \cdot B \cdot \gamma' + D' \gamma' \quad (3\text{-}9b)$
$\quad_{L/B=1}$

Plane Strain: $\quad \sigma'_{Izp} = B \cdot \gamma' + D \cdot \gamma' \quad (3\text{-}9c)$
$\quad_{L/B \geq 10}$

where σ'_{Izp} = effective overburden pressure at the depth of Izp in tsf; γ' = effective unit weight (wet soil unit weight γ less unit weight of water) in units of ton/ft³; and D = excavated or embedded depth in ft.

The parameter σ_{Izp} may be assumed to vary linearly between Equations 3-9b and 3-9c for L/B between 1 and 10. Iz may be assumed to vary linearly between 0.1 and 0.2 on the Iz axis at the ground surface for L/B between 1 and 10, and Z/B may be assumed to vary linearly between 2 and 4 on the Z/B axis for L/B between 1 and 10.

(2) Elastic Modulus. Elastic modulus E_{si} may be estimated from results of the mechanical (Dutch static) cone penetration test (item 59)

Axisymmetric footings: $\quad E_{si} = 2.5 \cdot q_c \quad (3\text{-}10a)$
$\quad_{L/B=1}$

Plane strain footings: $\quad E_{si} = 3.5 \cdot q_c \quad (3\text{-}10b)$
$\quad_{L/B \geq 10}$

where q_c is the cone-tip bearing resistance in tsf. E_{si} may be assumed to vary linearly between Equations 3-10a and 3-10b for L/B between 1 and 10. SPT data may also be converted to Dutch cone bearing capacity by the correlations in Table 3-1. The estimated average elastic modulus of each depth increment may be plotted in the E_s column of Figure 3-5.

(3) Calculation of Settlement. Iz/E_s is computed for each depth increment z/B and added to obtain SUM (Figure 3-5). Immediate settlement of the soil profile may then be calculated as shown in Figure 3-5. If a rigid base lies within $z = 2B$, then settlement may be calculated as shown down to the rigid base.

E. BURLAND AND BURBIDGE APPROXIMATION. This procedure, based on 200 SPT case studies, predicts settlements less than most of these methods (item 4).

(1) Immediate settlement of sand and gravel deposits may be estimated by

$$\Delta P'_{ave} > \sigma'_p: \quad \rho_i = f_s \cdot f_i \cdot \left[\left(\Delta P'_{ave} - \frac{2}{3} \sigma'_p \right) \cdot B^{0.7} \cdot I_c \right] \quad (3\text{-}11a)$$

$$\Delta P'_{ave} < \sigma'_p: \quad \rho_i = f_s \cdot f_i \cdot \Delta P'_{ave} \cdot \frac{I_c}{3} \quad (3\text{-}11b)$$

where f_s = shape correction factor, $[(1.25 \cdot L/B)/L/B + 0.25)]^2$; f_i = layer thickness correction factor, $H/z_1 \cdot (2 - H/z_1)$; $\Delta P'_{ave}$ = average effective bearing pressure, $q_{oave} + \sigma'_{oave}$ in tsf; q_{oave} = average pressure in stratum from foundation load in tsf; σ'_{oave} = average effective overburden pressure in stratum H in tsf; σ'_p = maximum effective past pressure in tsf; H = thickness of layer in ft; z_1 = depth of influence of loaded area in ft; I_c = compressibility influence factor, $\approx 0.23/(N^{1.4}_{ave})$

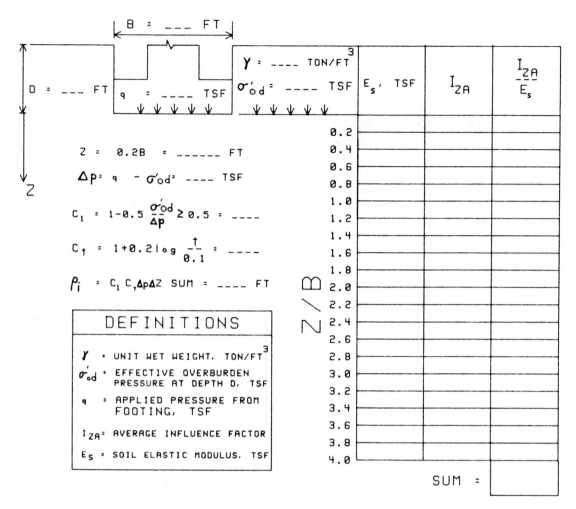

Figure 3-5. Settlement calculation sheet for cohesionless soil using Schmertmann's method

with coefficient of correlation 0.848; N_{ave} = average SPT blowcount over depth influenced by loaded area.

(a) The depth of influence z_1 is taken as the depth at which the settlement is 25 percent of the surface settlement. This depth in ft may be approximated by $1.35B^{0.75}$ where N_{ave} increases or is constant with depth. z_i is taken as $2B$ where N_{ave} shows a consistent decrease with depth.

Table 3-1. Correlations between Dutch Cone Tip Resistance q_c and Blow Count N from the SPT (Data from Item 55)

Soil (1)	q_c/N (2)
Silts, sandy silts, slightly cohesive silt-sand	2
Clean, fine to medium sands and slightly silty sands	3.5
Coarse sands and sands with little gravel	5
Sandy gravel and gravel	6

Note: Units of q_c are in tsf and N in blows/ft.

(b) N_{ave} is the arithmetic mean of the measured N values within the depth of influence z_i. N_{ave} is not corrected for effective overburden pressure, but instead considers compressibility using I_c. The arithmetic mean of the measured N_{ave} should be corrected to $15 + 0.5(N_{ave} - 15)$ when $N_{ave} > 15$ for very fine and silty sand below the water table and multiplied by 1.25 for gravel or sandy gravel.

(c) The probable limits of accuracy of Equations 3-11 are within upper- and lower-bound values of I_c given by

$$0.08/(N_{ave})^{1.3} \le I_c \le 1.34/N_{ave})^{1.67} \quad (3\text{-}12)$$

(2) Settlement after time t at least three years following construction from creep and secondary compression effects may be estimated by

$$\rho_t = f_t \cdot \rho_i \quad (3\text{-}13)$$

where $f_t = 1 + R_3 + R_t \cdot \log t/3$; R_3 = time-dependent settlement ratio as a proportion of ρ_i during the first three years following construction, ≈ 0.3; and $R_t =$ time-dependent settlement ratio as a proportion of ρ_i for each log cycle of time after three years, ≈ 0.2. Values of R_3 and R_t are conservative based on nine case records (item 4).

F. DILATOMETER APPROXIMATION. The dilatometer consists of a stainless-steel blade 96 mm wide and 15 mm thick with a sharp edge containing a stainless-steel membrane centered and flush with one side of the blade. The blade is preferably pushed (or driven if necessary) into the soil. A pressure-vacuum system is used to inflate/deflate the membrane a maximum movement of 1.1 mm against the adjacent soil (item 58).

(1) Calculation. This procedure predicts settlement from evaluation of one-dimensional vertical compression or constrained modulus E_d by the DMT

$$\rho_i = \frac{q_{oave} \cdot H}{E_d} \qquad (3\text{-}14)$$

where q_{oave} = average increase in stress caused by the applied load in tsf; H = thickness of stratum at depth z where q_{oave} is applicable in ft; E_d = constrained modulus, $R_D E_s$, in tsf; $R_D = (1 - \nu_s)/[(1 + \nu_s)(1 - 2\nu_s)]$, a factor that varies from 1 to 3 relates E_d to Young's soil modulus E_s; and ν_s = Poisson's ratio.

Refer to Appendix D for additional information on elastic parameters. The influence of prestress on settlement may be corrected using results of DMT and CPT tests after Schmertmann's approximation (item 37) to reduce settlement overestimates.

(2) Evaluation of Elastic Modulus. The dilatometer modulus of soil at the depth of the probe is evaluated as 34.7 times the difference in pressure between the deflated and inflated positions of the membrane. Young's elastic modulus has been found to vary from 0.4 to 10 times the dilatometer modulus (item 39). A Young's elastic modulus equal to the dilatometer modulus may be assumed for many practical applications in sands.

(3) Adjustment for Other Soil. The constrained modulus E_d may be adjusted for effective vertical stress σ_o' other than that of the DMT for overconsolidated soil and normally consolidated clay by

$$E_d = m \cdot \sigma_o' \qquad (3\text{-}15a)$$

where $m = [(1 + e)/C_c] \cdot \ln 10$; e = void ratio; and C_c = compression index. The constrained modulus for normally consolidated silts and sand is

$$E_d = m \cdot (\sigma_o')^{0.5} \qquad (3\text{-}15b)$$

where σ_o' is the effective vertical overburden pressure in tsf. These settlements include time-dependent settlements excluding secondary compression and creep. Total settlement of a heterogeneous soil with variable E_d may be estimated by summing increments of settlement using Equation 3-14 for layers of thickness H.

3-4. Recommendations

A minimum of three methods should be applied to estimate a range of settlement. Settlement estimates based on in situ test results are based on correlations obtained from past experience and observation and may not be reliable.

A. EVALUATION FROM SPT DATA. The Alpan (Equation 3-2), Schultze and Sherif (Equation 3-4), and modified Terzaghi and Peck (Equation 3-5) approximations should all be applied to estimate immediate settlement if blowcount data from SPT are available. The Burland and Burbidge approximation (Equations 3-11) should be applied if the maximum past pressure of the soil can be estimated; this approximation using Equation 3-12 may also be applied to estimate a range of settlement.

B. EVALUATION FROM CPT DATA. The Schmertmann approximation (Equation 3-8) should be used to estimate settlement if CPT data are available.

C. EVALUATION FROM DMT DATA. The dilatometer approximation (Equation 3-14) should be used if data from this test are available. The range of settlement may be determined by assuming minimum and maximum values of the factor R_D of 1 and 3.

D. EVALUATION FROM PMT DATA. The pressuremeter unload-reload modulus from the corrected pressure versus volume change curve is a measure of twice the shear modulus (Appendix D-2d). Young's elastic modulus may be evaluated from the shear modulus (Table D-2), and settlement can be estimated from Equation 3-8. The constrained modulus may be evaluated from Young's elastic modulus (Table D-2), and settlement can be estimated from Equation 3-14.

E. LONG-TERM SETTLEMENT. The Schmertmann and Burland and Burbidge approximations may be used to estimate long-term settlement in cohesionless soil from CPT and SPT data. The constrained modulus E_d may also be adjusted to consider consolidation from Equations 3-15 and settlement estimated from Equation 3-14. Refer to items 39 and 58 for detailed information on evaluation of the constrained modulus.

3-5. Application

A footing 10-ft square is to be constructed 3 ft below grade on medium dense ($\gamma = \gamma' = 0.06$ ton/ft³) and moist sand with total stratum thickness of 13 ft ($H = 10$ ft). The water table is at least 5 ft below the base of the footing. The effective vertical overburden pressure at the bottom of the footing is $\sigma'_{od} = \gamma' \cdot z = 0.06 \cdot 3 = 0.18$ tsf. The bearing pressure of the footing on the sand $q = 2$ tsf. Field data indicate an average blowcount in the sand $N_{ave} = 20$ blows/ft and the cone-tip-bearing resistance is about 70 tsf. The average elastic modulus determined from dilatometer and pressuremeter tests indicated $E_s = 175$ tsf. Refer to Figure 3-6 for a schematic description of this problem. Estimates of settlement of this footing at the end of construction (EOC) and 10 years after construction are required.

(1) Results of the settlement computations comparing several of these methods are shown in Tables 3-2a and 3-2b.

(a) Figure 3-6 shows computation of settlement by Schmertmann's method.

(b) Computation of settlement by the Burland and Burbidge and dilatometer approximations requires an estimate of the average effective bearing pressure $\Delta P'_{ave}$. Assuming that the 2:1 stress distribution of Figure C-1 is adequate, the average pressure from the foundation load is

$$q_{oave} = \frac{q + \Delta\sigma_z}{2}$$

where $\Delta\sigma_z$ is found from Equation C-2. Therefore, if $B = L = H = 10$ ft and $Q = q \cdot B \cdot L$, then

$$q_{oave} = 0.5 \left[2.0 + \frac{2.0 \cdot 10^2}{(10 + 10)^2} \right] = 1.25 \text{ tsf}$$

The average effective overburden pressure $\sigma'_{oave} = 0.06 \cdot (3 + 13)/2.0$ or 0.48 tsf. The average effective bear-

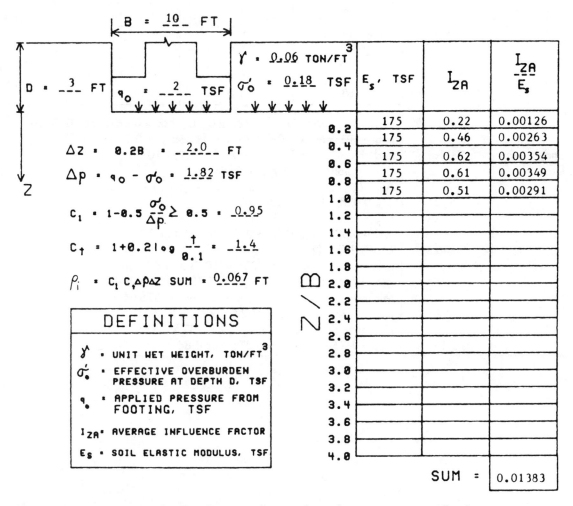

Figure 3-6. Estimation of immediate settlement by Schmertmann's method

Table 3-2a. Estimation of Immediate Settlement for Example Application of Footing on Cohesionless Soil: Calculations

Method (1)	Equation (2)	Calculations (3)
Alpan (item 1)	—	$B = 10$ ft, $L/B = 1$, $m' = 1.0$, $N' = 65$ blows/ft
	—	$N = 20$ blows/ft and $\sigma'_o = 0.18$ tsf (Figure 3-1a)
	—	$\alpha_o = 0.05$ in./tsf (Figure 3-1b)
	3-2	$\rho_i = 1 \cdot \left[\dfrac{2 \cdot 10}{1 + 10} \right]^2 \cdot \dfrac{0.05}{12} \cdot 2 = 0.027$ ft or 0.33 in.
Schultz and Sherif (item 60)	—	$H/B = 1$, $D/B = 0.3$, $f = 0.052$ (Figure 3-2)
	3-4	$\rho_i = \dfrac{0.052 \cdot 2 \cdot \sqrt{10}}{20^{0.87} (1 + 0.4 \cdot 0.3)} = 0.022$ ft or 0.33 in.
Modified Terzaghi and Peck (item 51)	—	$C_n = 1.6$ (Figure 3-3b)
	—	$C_w = 1$
	3-6	$N' = 1 \cdot 1.6 \cdot 20 = 32$
	—	$q_1 = 3.5$ tsf (Figure 3-3a)
	3-5	$\rho_i = (1/18) \cdot (2/3.5) = 0.031$ ft or 0.38 in. (0.57 in. ignoring Peck and Bazarra 1/3 reduction)
Schmertmann (item 55)	—	Refer to Figure 3-6
	3-10a	$q_c = 70$ tsf
	3-9b	$E_{si} = 2.5 \cdot 70 = 175$ tsf
	3-9a	$\sigma'_{lzp} = 0.5 \cdot 10 \cdot 0.06 + 3 \cdot 0.6 = 0.48$ tsf
	—	$Izp = 0.5 + 0.1 (1.82/0.48)^{0.5} = 0.695$

z (ft)	z/B	Iz	Iza
0	0.0	0.10	0.22
2	0.2	0.34	0.46
4	0.4	0.58	0.62
6	0.6	0.65	0.61
8	0.8	0.56	0.51
10	1.0	0.46	

Method (1)	Equation (2)	Calculations (3)
	3-8	ρ_i at EOC $= 0.95 \cdot 1.0 \cdot 1.82 \cdot 2.0 \cdot 0.0138 = 0.0477$ ft or 0.57 in.
	—	ρ_i after 10 years $= 0.0477 \cdot 1.4 = 0.0668$ ft or 0.80 in.
Burland and Burbidge (item 4)	3-11a	End of $\rho_i = 1 \cdot 0.91[(1.73 - 0.67 \cdot 0.18) \cdot 10^{0.7} \cdot 0.0035]$
	—	Construction: $= 0.028$ ft or 0.34 in.
	—	After 10 years: $f_t = 1 + 0.3 + 0.2 \log (10/3) = 1.4$
	3-13	$\rho_t = 0.028 \cdot 1.4 = 0.039$ or 0.47 in.
	3-12	$I_{cmin} \approx 0.08/(20)^{1.3} = 0.0016$
	—	$\rho_{imin} = 0.028 \cdot 0.0016/0.0035 = 0.013$ ft or 0.15 in.
	3-12	$I_{cmax} \approx 1/34/(20)^{1.67} = 0.009$
	—	$\rho_{imax} = 0.028 \cdot 0.009/0.0035 = 0.073$ ft or 0.87 in.
Dilatometer (item 58)	—	$q_{oave} = 1.25$ tsf, $H = 10$ ft, $E_s = 175$ tsf, ν_s not needed
	—	Minimum settlement: $R_D = 3$, $E_d = 3 \cdot 175 = 525$ tsf
	3-14	$\rho_{imin} = 1.25 \cdot 10/525 = 0.024$ ft or 0.29 in.
	—	Maximum settlement: $R_D = 1$, $E_d = 1 \cdot 175 = 175$ tsf
	3-14	$\rho_{imax} = 1.25 \cdot 10/175 = 0.071$ ft or 0.86 in.

Table 3-2b. Comparison

Method (1)	Immediate settlement, ft (in.)	
Alpan	0.027	(0.33)
Schultz and Sherif	0.022	(0.27)
Modified Terzaghi and Peck	0.031	(0.38)
Schmertmann	0.048	(0.57)
Burland and Burbidge	0.028	(0.34)
Dilatometer	0.024 to 0.071	(0.29 to 0.86)

ing pressure $\Delta P'_{ave}$ is therefore $1.25 + 0.48 = 1.73$ tsf. The soil is assumed to be normally consolidated; therefore, $\sigma'_p = \sigma'_{od} = 0.18$ tsf, and Equation 3-11a is applicable. Factor $f_s = 1.0$, $H/z_1 = 1.31$, and $f_1 = 0.91$. $I_c = 0.23/(20)^{1.4} = 0.0035$.

(2) A comparison of results in Table 3-2b shows that the Alpan, Schultze and Sherif, modified Terzaghi and Peck, and Burland and Burbidge methods provide consistent settlements of about 0.3 to 0.4 in. The Schmertmann method is reasonably conservative with settlement of 0.57 in. This settlement is the same as that from the modified Terzaghi and Peck method, ignoring the 1/3 reduction recommended by Peck and Bazarra (item 50). Long-term settlement is 0.5 (Burland and Burbidge) and 0.8 in. (Schmertmann) after 10 years. The expected range of settlement is 0.2 to 1.0 in. after the Burland and Burbidge method and 0.3 to 0.9 in. from the dilatometer. Settlement is not expected to exceed 1 in.

Section II. Immediate Settlement of Cohesive Soil for Static Loads

3-6. General

Static loads cause immediate and long-term consolidation settlements in cohesive or compressible soil. The stress in the soil caused by applied loads should be estimated (paragraph 1-5d) and compared with estimates of the maximum past pressure (paragraph 1-5a). If the stress in the soil exceeds the maximum past pressure, then primary consolidation and secondary compression settlement may be significant and should be evaluated by the methods in Sections III and IV. Immediate rebound or heave may occur in compressible soil at the bottom of excavations, but may not be a design or construction problem unless rebound causes the elevation of the basement or first floor to exceed specifications or impair performance.

3-7. Rebound in Excavations

Most rebound in excavations lying above compressible strata occurs from undrained elastic unload-

ing strains in these strata. Additional long-term heave due to wetting of the soil following reduction in pore water pressure following removal of overburden in excavations is discussed in Chapter 5, Section I. Rebound of compressible soil in excavations may be approximated as linear elastic by (item 2)

$$S_{RE} = F_{RD} \cdot F_{RS} \cdot \frac{\gamma D^2}{E_s^*} \qquad (3-16)$$

where S_{RE} = undrained elastic rebound in ft; F_{RD} = rebound depth factor (Figure 3-7a); F_{RS} = rebound shape

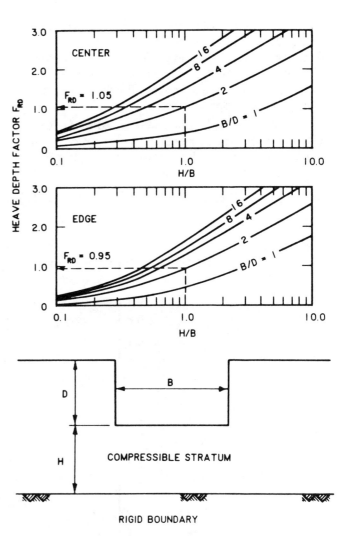

a. REBOUND DEPTH FACTOR F_{RD}

Figure 3-7. Factors to calculate elastic rebound in excavations. Reprinted by permission of the author G. Y. Baladi from "Distribution of Stresses and Displacements Within and Under Long Elastic and Viscoelastic Embankments," Ph.D. Thesis, 1968, Purdue University

factor (Figure 3-7b); γ = wet unit weight of excavated soil in tons/ft^3; D = depth of excavation in ft; and E_s^* = equivalent elastic modulus of soil beneath the excavation in tsf.

The equivalent elastic modulus E_s^* may be estimated by methods described in Appendix D, Elastic Parameters. The compressible stratum of depth H is assumed to be supported on a rigid base such as unweathered clay shale, rock, dense sand, or gravel. An example application is provided in Figure 3-7c.

3-8. Immediate Settlement in Cohesive Soil

The immediate settlement of a structure on cohesive soil (see paragraph 1-5c for definition) consists of elastic distortion associated with a change in shape without volume change and, in unsaturated clay, settlement from a decrease in volume. The theory of elasticity is generally applicable to cohesive soil.

A. IMPROVED JANBU APPROXIMATION. The average immediate settlement of a foundation on an elastic soil may be given by (item 9)

$$\rho_i = \mu_o \cdot \mu_1 \cdot \frac{q \cdot B}{E_s^*} \qquad (3\text{-}17)$$

where μ_o = influence factor for depth D of foundation below ground surface (Figure 3-8); μ_1 = influence factor for foundation shape (Figure 3-8); and E_s^* = equivalent Young's modulus of the soil in tsf.

(1) A comparison of test calculations and results of finite element analysis have indicated errors from Equation 3-17 usually less than 10% and always less than 20% for H/B between 0.3 and 10, L/B between 1 and 5, and D/B between 0.3 and 3 (Figure 3-8). Reasonable results are given in most cases when μ_o is set equal to unity. Poisson's ratio v_s is taken as 0.5.

(2) E_s^* may be estimated by methods given in Appendix D.

B. PERLOFF APPROXIMATION. The immediate vertical settlement beneath the center and edge of a mat or footing may be given by (item 52)

$$\rho_i = I \cdot q \cdot B \cdot \left[\frac{1 - v_s^2}{E_s} \right] \cdot \alpha \qquad (3\text{-}18)$$

where I = influence factor for infinitely deep and homogeneous soil (Table 3-3a); E_s = elastic soil modulus in tsf; v_s = soil Poisson's ratio; and α = correction factor for subgrade soil (Table 3-3b).

The influence factor I may be modified to ac-

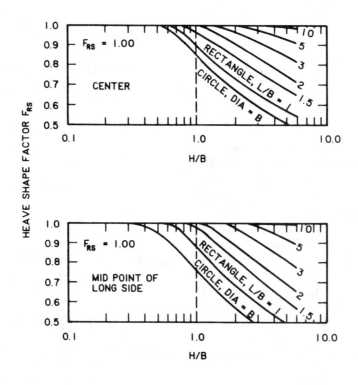

b. REBOUND SHAPE FACTOR F_{RS}

Figure 3-7. (Continued)

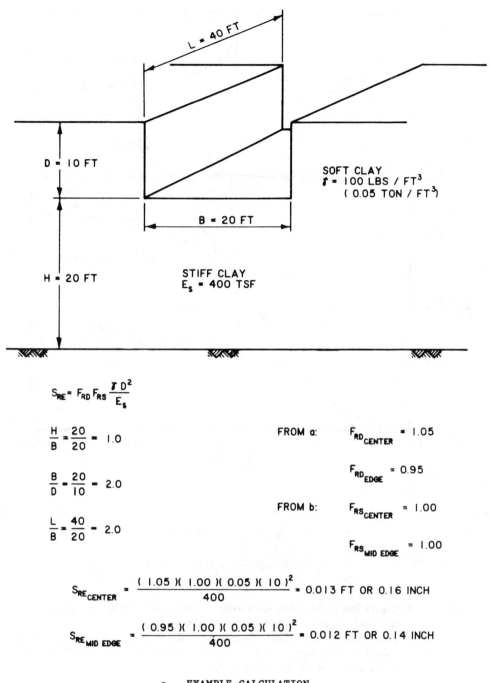

$$S_{RE} = F_{RD} F_{RS} \frac{\gamma D^2}{E_s}$$

$$\frac{H}{B} = \frac{20}{20} = 1.0$$

FROM a: $F_{RD_{CENTER}} = 1.05$

$$\frac{B}{D} = \frac{20}{10} = 2.0$$

$F_{RD_{EDGE}} = 0.95$

$$\frac{L}{B} = \frac{40}{20} = 2.0$$

FROM b: $F_{RS_{CENTER}} = 1.00$

$F_{RS_{MID\ EDGE}} = 1.00$

$$S_{RE_{CENTER}} = \frac{(1.05)(1.00)(0.05)(10)^2}{400} = 0.013 \text{ FT OR } 0.16 \text{ INCH}$$

$$S_{RE_{MID\ EDGE}} = \frac{(0.95)(1.00)(0.05)(10)^2}{400} = 0.012 \text{ FT OR } 0.14 \text{ INCH}$$

c. EXAMPLE CALCULATION

Figure 3-7. (Concluded)

count for heterogeneous or multilayered soil usually encountered in practice. If the upper soil is relatively compressible and underlain by stiff clay, shale, rock, or dense soil, then the compressible soil stratum may be approximated by a finite layer of depth H supported on a rigid base. The influence factor I is given in Figure 3-9 for settlement beneath the center and midpoint of the edge of flexible foundations. If the subgrade soil supporting the foundation with modulus E_{s1} and thickness H is underlain by less rigid infinitely deep material with modulus E_{s2}, then settlement at the center of a uniformly loaded circular area placed on the surface of the more rigid soil is corrected with the factor (Table 3-3b).

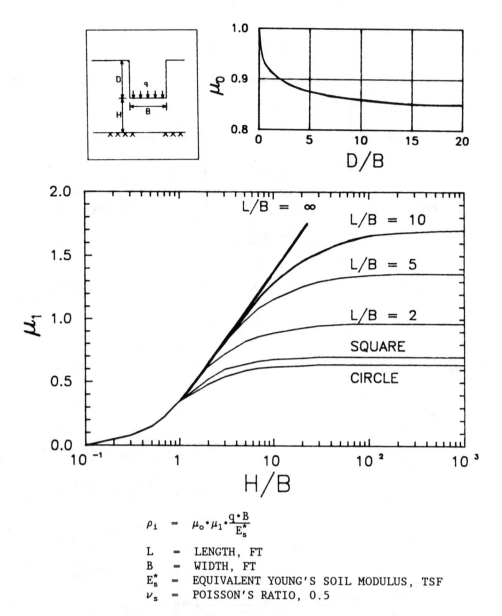

$$\rho_i = \mu_o \cdot \mu_1 \cdot \frac{q \cdot B}{E_s^*}$$

L = LENGTH, FT
B = WIDTH, FT
E_s^* = EQUIVALENT YOUNG'S SOIL MODULUS, TSF
ν_s = POISSON'S RATIO, 0.5

Figure 3-8. Chart for estimating immediate settlement in cohesive soil. Reprinted by permission of the National Research Council of Canada from *Canadian Geotechnical Journal*, Vol 15, 1978, "Janbu, Bjerrum, and Kjaernsli's Chart Reinterpreted", by J. T. Christian and W. D. Carrier III, p. 127

C. KAY AND CAVAGNARO APPROXIMATION. The immediate elastic settlement at the center and edge of circular foundations and foundations with length to width ratios less than 2 may be evaluated for layered elastic soil by the graphical procedure (Figure 3-10) (item 31). The method considers the relative rigidity of the foundation relative to the soil and can evaluate the differential displacement between the center and edge of the foundation.

3-9. Recommendations

A. JANBU APPROXIMATION. The Janbu approximation is recommended when an average computation of settlement is required for a wide range of depths, lengths, and widths of foundations supported on compressible soil of depth *H*.

B. PERLOFF APPROXIMATION. The Perloff approximation should be used when total and

Table 3-3a. Factors for Estimating Immediate Settlement in Cohesive Soil: Shape and Rigidity Factor I for Calculating Settlements of Points on Loaded Areas at the Surface of an Elastic Half-Space (Data from Item 52)

Shape (1)	Length/Width (2)	Center center (3)	Corner corner (4)	Middle short side (5)	Middle long side (6)
Circle	—	1.00	0.64	0.64	0.64
Rigid circle	—	0.79	—	—	—
Square	—	1.12	0.56	0.76	0.76
Rigid square	—	0.99	—	—	—
Rectangle	1.5	1.36	0.67	0.89	0.97
Rectangle	2	1.52	0.76	0.98	1.12
Rectangle	3	1.78	0.88	1.11	1.35
Rectangle	5	2.10	1.05	1.27	1.68
Rectangle	10	2.53	1.26	1.49	2.12
Rectangle	100	4.00	2.00	2.20	3.60
Rectangle	1,000	5.47	2.75	2.94	5.03
Rectangle	10,000	6.90	3.50	3.70	6.50

Table 3-3b. Factors for Estimating Immediate Settlement in Cohesive Soil: Correction Factor α at the Center of a Circular Uniformly Loaded Area of Width B on an Elastic Layer of Modulus E_{s1} of Depth H Underlain by a Less Stiff Elastic Material of Modulus E_{s2} of Infinite Depth

H/B (1)	E_{s1}/E_{s2}				
	1 (2)	2 (3)	5 (4)	10 (5)	100 (6)
0	1.000	1.000	1.000	1.000	1.000
0.1	1.000	0.972	0.943	0.923	0.760
0.25	1.000	0.885	0.779	0.699	0.431
0.5	1.000	0.747	0.566	0.463	0.228
1.0	1.000	0.627	0.399	0.287	0.121
2.5	1.000	0.550	0.274	0.175	0.058
5	1.000	0.525	0.238	0.136	0.036
∞	1.000	0.500	0.200	0.100	0.010

Reprinted from D.M. Burmister, 1965, "Influence Diagrams for Stresses and Displacements in a Two-Layer Pavement System for Airfields," Contract NBY 13009, Department of the Navy, Washington, D.C. (item 7).

differential settlement is required beneath flexible foundations located at or near the surface of the soil; settlements may be evaluated at the center, corner, and middle edges of both the short and long sides of the foundation.

C. KAY AND CAVAGNARO APPROXIMATION. The Kay and Cavagnaro approximation should be used when total and differential settlement is required beneath footings and mats of a given stiffness supported on compressible soil of variable elastic modulus; settlement may be evaluated at the center and edge for a given foundation depth. A reasonable estimate of Poisson's ratio for cohesive soil is 0.4 (Appendix D-4).

D. LINEAR MODULUS INCREASE. The Gibson model described in Appendix D-2d may be used if the elastic modulus may be assumed zero at the ground surface. A parametric analysis using the Kay and Cavagnaro graphical procedure for an elastic modulus increasing linearly with depth indicates that the center settlement beneath a foundation may be calculated by

$$\rho_c = \frac{q}{k}[0.7 + (2.3 - 4v_s)\log_{10}n] \quad (3\text{-}19)$$

where $n = kR/(E_o + kD_b)$; k = constant relating the elastic modulus with depth, i.e., $E_o = E_s + kz$ in ksf/ft;

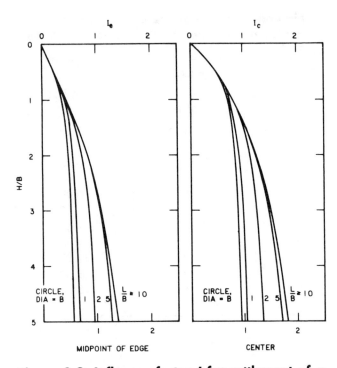

Figure 3-9. Influence factor _I_ for settlement of a completely flexible mat or footing of width _B_ and length _L_ on a finite elastic material of depth _H_ supported on a rigid base. Data taken with permission of McGraw-Hill Book Company from Tables 2-4 and 2-5, _Foundations of Theoretical Soil Mechanics_, by M. E. Harr, 1966, p. 98, 99

R = equivalent radius of the mat or footing, $\sqrt{LB/\pi}$; E_o = elastic soil modulus at the ground surface in ksf; and D_b = depth of the mat base or stiffening beams beneath the ground surface in ft.

Edge and corner settlement of a flexible mat or footing will be approximately 1/2 and 1/4 of the center settlement, respectively. Differential movement of the mat or footing may be calculated from Figure 3-10.

3-10. Application

A footing 10 ft square and 1 ft thick with base 3 ft below ground surface, is to be constructed on cohesive soil. The pressure applied on the footing is q = 2 tsf (4 ksf). The equivalent elastic modulus of this clay, which is 10 ft deep beneath the footing, is 175 tsf (350 ksf) and Poisson's ratio is 0.4. Table 3-4 compares settlement computed by the improved Janbu and Perloff methods. Refer to Figure 3-11 for application of the Kay and Cavagnaro method.

a. Average settlement by the improved Janbu method is 0.48 in.

b. The Kay and Cavagnaro method in Figure 3-11 calculates smaller edge settlement of 0.33 in. compared with 0.46 in. and smaller center settlement of 0.73 in. compared with 0.81 in. calculated from the Perloff method. Actual differential settlement when considering stiffness of the footing is only about 0.02 in. (Figure 3-11); the footing is essentially rigid. Settlement will be less than 1 in. and is expected to be about 0.5 in.

Section III. Primary Consolidation Settlement

3-11. Description

Vertical pressure σ_{st} from foundation loads transmitted to a saturated compressible soil mass is initially carried by fluid or water in the pores because water is relatively incompressible compared with that of the soil structure. The pore water pressure u_{we} induced in the soil by the foundation loads is initially equal to the vertical pressure σ_{st}; it is defined as excess pore water pressure because this pressure exceeds that caused by the weight of water in the pores. Primary consolidation begins when water starts to drain from the pores. The excess pressure and its gradient decrease with time as water drains from the soil, causing the load to be gradually carried by the soil skeleton. This load transfer is accompanied by a decrease in volume of the soil mass equal to the volume of water drained from the soil. Primary consolidation is complete when all excess pressure has dissipated so that $u_{we} = 0$ and the increase in effective vertical stress in the soil $\Delta\sigma' = \sigma_{st}$. Primary consolidation settlement is usually determined from results of one-dimensional consolidometer tests. Refer to Appendix E for a description of 1-D consolidometer tests.

A. NORMALLY CONSOLIDATED SOIL. A normally consolidated soil is a soil subject to an in situ effective vertical overburden stress σ_o' equal to the preconsolidation stress σ_p'. Virgin consolidation settlement for applied stresses exceeding σ_p' can be significant in soft and compressible soil with a skeleton of low elastic modulus such as plastic CH and CL clays, silts, and organic MH and ML soils.

B. OVERCONSOLIDATED SOIL. An overconsolidated soil is a soil that is subject to an in situ effective overburden stress σ_o' less than σ_p'. Consolidation settlement will be limited to recompression from stresses applied to the soil up to σ_p'. Recompression settlement is usually much less than virgin consolidation settlement caused by applied stresses exceeding σ_p'.

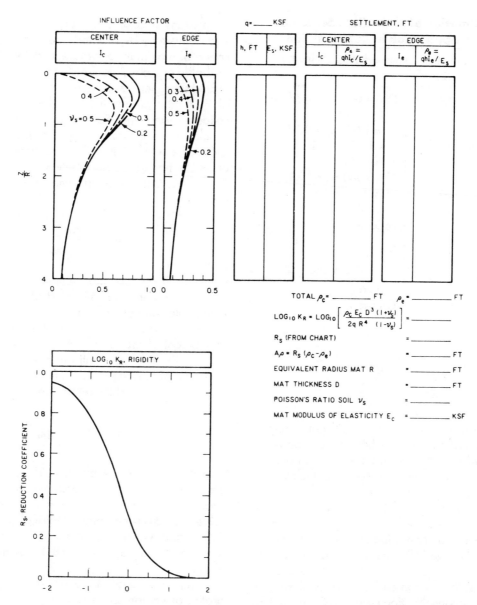

Figure 3-10. Computation of elastic settlement beneath a mat foundation (data from item 31); z = depth beneath mat, ft; R = equivalent radius $\sqrt{LB/\pi}$ in ft

3-12. Ultimate 1-D Consolidation

The ultimate or long-term 1-D consolidation settlement is initially determined followed by adjustment for overconsolidation effects. Refer to Table 3-5 for the general procedure to determine ultimate settlement by primary consolidation.

A. EVALUATION OF VOID RATIO-PRESSURE RELATIONSHIP. Estimates of the ultimate consolidation settlement following complete dissipation of hydrostatic excess pressure requires determination of the relationship between the in situ void ratio and effective vertical stress in the soil. The

loading history of a test specimen taken from an undisturbed and saturated soil sample, for example, may be characterized by a void-ratio-versus-logarithm pressure diagram (Figure 3-12).

(1) Correction of Laboratory Consolidation Curve. Removal of an impervious soil sample from its field location will reduce the confining pressure, but tendency of the sample to expand is restricted by the decrease in pore water pressure. The void ratio will tend to remain constant at constant water content because the decrease in confinement is approximately balanced by the decrease in water pressure; therefore, the effective stress remains constant in theory after

Table 3-4. Estimation of Immediate Settlement for Example Application in Cohesive Soil

Method (1)	Equation (2)	Immediate settlement, ρ_i (3)
Janbu and Senneset (1982) (item 9)	3-17	$D/B = 0.3$, $H/B = L/B = 1.0$ $\mu_o = 1.00$, $\mu_1 = 0.35$ $\rho_i = 1.000 \cdot 352 \cdot 0 \cdot 10/175$ $= 0.040$ ft or 0.48 in.
Perloff (1975) (item 52)	3-18	From Figure 3-9, $I_e = 0.4$, $I_c = 0.7$ Edge: $\rho_i = 0.4 \cdot 2 \cdot 10 \cdot \dfrac{1 - 0.16}{175}$ $= 0.038$ ft or 0.46 in. Center: $\rho_i = 0.7 \cdot 2.0 \cdot 10 \cdot \dfrac{1 - 0.16}{175}$ $= 0.067$ ft or 0.81 in.

Equation 1-1 and the void ratio should not change. Classical consolidation assumes that elastic expansion is negligible and the effective stress is constant during release of the in situ confining pressure after the sample is taken from the field. Some sample disturbance occurs, however, so that the laboratory consolidation curve must be corrected as shown in Figure 3-12. Perfectly undisturbed soil should indicate a consolidation curve similar to line e_oED (Figure 3-12a) or line e_oBFE (Figure 3-12b). Soil disturbance increases the slope for stresses less than the preconsolidation stress illustrated by the laboratory consolidation curves in Figure 3-12. Pushing undisturbed samples into metal Shelby tubes and testing in the consolidometer without removing the horizontal restraint help maintain the in situ horizontal confining pressure, reduce any potential volume change following removal from the field, and help reduce the correction for sample disturbance.

(2) Normally Consolidated Soil. A normally consolidated soil in situ will be at void ratio e_o and effective overburden pressure σ'_o equal to the preconsolidation stress σ'_p. e_o may be estimated as the initial void ratio prior to the test if the water content of the sample did not change during storage and soil expansion is negligible. In situ settlement from applied loads is determined from the field virgin consolidation curve.

(a) Reconstruction of the field virgin consolidation curve with slope C_c shown in Figure 3-12a may be estimated by the procedure in Table 3-6a. Determining the point of greatest curvature for evaluation of

the preconsolidation stress requires care and judgment. Two points may be selected bounding the probable location of maximum curvature to determine a range of probable preconsolidation stress. Higher-quality undisturbed specimens assist in reducing the probable range of σ'_p. If σ'_p is greater than σ'_o, then the soil is overconsolidated and the field virgin consolidation curve should be reconstructed by the procedure in Table 3-6b. The scale of the plot may have some influence on evaluation of the parameters.

(b) Consolidation settlement may be estimated by

$$\rho_{cj} = \frac{\Delta e_j}{1 + e_{oj}} \cdot H_j \qquad (3\text{-}20)$$

where ρ_{cj} = consolidation settlement of stratum j in ft; Δe_j = change in void ratio of stratum j, $e_{oj} - e_{fj}$; e_{oj} = initial void ratio of stratum j at initial pressure σ'_{oj}; e_{fj} = final void ratio of stratum j at final pressure σ'_{fj}; and H_j = height of stratum j in ft.

The final void ratio may be found graphically using the final pressure σ'_f illustrated in Figure 3-12a. The change in void ratio may be calculated by

$$\Delta e_j = C_c \cdot \log_{10} \frac{\sigma'_{fj}}{\sigma'_{oj}} \qquad (3\text{-}21)$$

where C_c is the slope of the field virgin consolidation curve or compression index. Figure 3-13 illustrates evaluation of C from results of a 1-D consolidation test. Table 3-7 illustrates some empirical correlations of C_c with natural water content, void ratio, and liquid limit. Refer to chapter 3, TM 5-818-1, for further estimates of C_c.

(c) Total consolidation settlement ρ_c of the entire profile of compressible soil may be determined from the sum of the settlement of each stratum

$$\rho_c = \sum_{j=1}^{n} \rho_{cj} \qquad (3\text{-}22)$$

where n is the total number of compressible strata. This settlement is considered to include much of the immediate elastic compression settlement ρ_i (Equation 3-1).

(3) Apparent Preconsolidation. A presumably normally consolidated soil may exhibit an apparent preconsolidation stress σ'_{qp} (Figure 3-12a). σ'_{qp} may be caused by several mechanisms; for example, the most common cause is secondary compression or the gradual reduction in void ratio (accompanied by an increase in attractive force between particles) at constant effective stress over a long time. Other causes of σ'_{qp} include a change in pore fluid, which causes attractive forces between particles to in-

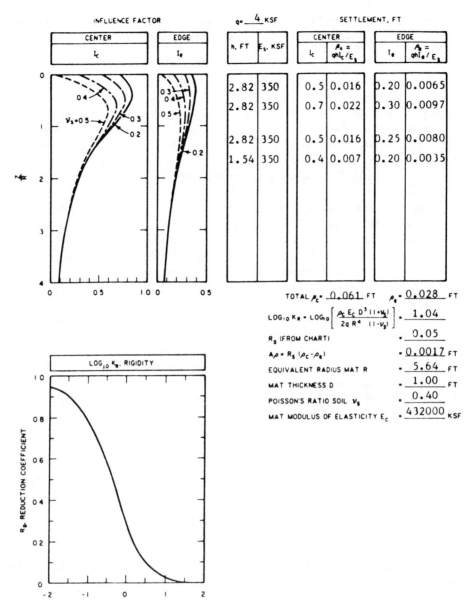

Figure 3-11. Estimation of immediate settlement for the example problem by the Kay and Cavagnaro method

crease, or cementation due to precipitation of cementatious materials from flowing groundwater. This apparent preconsolidation is sensitive to strain and may not be detected because of sample disturbance. Existence of σ'_{qp} in the field can substantially reduce settlement for a given load and can be used to reduce the factor of safety or permit greater pressures to be placed on the foundation soil, provided that collapse will not be a problem. Refer to Chapter 5, Sections 5-7 to 5-10 for estimating potential collapse.

(4) Overconsolidated Soil. An overconsolidated soil will be at a void ratio e_o and effective vertical confining pressure σ'_o represented by point B (Figure 3-12b). At some time in the past the soil was subject to the preconsolidation stress σ'_p, but this pressure was later reduced, perhaps by soil erosion or removal of glacial ice, to the existing overburden pressure σ'_o. The in situ settlement for an applied load will be the sum of recompression settlement between points B and F and any virgin consolidation from a final effective vertical applied pressure σ'_f exceeding the preconsolidation stress σ'_p. Reloading a specimen in the consolidometer will give the laboratory curve shown in Figure 3-12b.

(a) Reconstruction of the field virgin consolidation curve with slope C_c may be estimated by the procedure in Table 3-6b. Refer to Table 3-7 for methods of estimating C_c.

Table 3-5. Procedure for Calculation of Ultimate Primary Consolidation Settlement of a Compressible Stratum

Step (1)	Description (2)
1	Evaluate the preconsolidation stress σ'_p from results of a one-dimensional consolidation test on undisturbed soil specimens using the Casagrande construction procedure (Table 3-6a) or by methods described in paragraph 1-5a. Refer to Appendix E for a description of 1-D consolidation tests.
2	Estimate the average initial effective overburden pressure σ'_o in each compressible stratum using soil unit weights, depth of overburden on the compressible stratum, and the known groundwater level or given initial pore water pressure in the stratum. Refer to Equation 1-1, $\sigma'_{oz} = \gamma z - u_w$. $\sigma'_o = (\sigma'_{oz1} + \sigma'_{oz2})/2$ where σ'_{oz1} = effective pressure at top of compressible stratum and σ'_{oz2} = effective pressure at the bottom of the compressible stratum.
3	Determine the soil initial void ratio e_o as part of the 1-D consolidation test or by methods given in Appendix II, EM 1110-2-1906, Laboratory Soils Testing.
4	Evaluate the compression index C_c from results of a 1-D consolidation test using the slope of the field virgin consolidation line determined by the procedure in Table 3-6a as shown in Figures 3-12 and 3-13, or preliminary estimates may be made from Table 3-7. Determine the recompression index C_r for an overconsolidated soil as illustrated in Figures 3-12 and 3-13; preliminary estimates may be made from Figure 3-14.
5	Estimate the final applied effective pressure σ'_f where $\sigma'_f = \sigma'_o + \sigma_{st}$. σ_{st}, soil pressure caused by the structure, may be found from Equation C-2 or the Boussinesq solution in Table C-1.
6	Determine the change in void ratio Δe_j of stratum j for the pressure increment $\sigma'_f - \sigma'_o$ graphically from a data plot similar to Figure 3-12, from Equation 3-21 for a normally consolidated soil, or from Equation 3-23 for an overconsolidated soil.
7	Determine the ultimate one-dimensional consolidation settlement of stratum j with thickness H_j, from Equation 3-20 $$\rho_{cj} = \frac{\Delta e_j}{1 + e_{oj}} H_j$$
8	Determine the total consolidation ρ_c of the entire profile of compressible soil from the sum of the settlement of each stratum (Equation 3-22) $$\rho_c = \sum_{j=1}^{n} \rho_{cj}$$
9	Correct ρ_c for effect of overconsolidation and small departures from 1-D compression on the initial excess pore pressure using the Skempton and Bjerrum procedure (Equation 3-24) $$\rho_{\lambda c} = \lambda \rho_c$$ where λ is found from Figure 3-15. $\lambda = 1$ if $B/H > 4$ or if depth to the compressible stratum is $> 2B$. The equivalent dimension of the structure when corrected to the top of the compressible stratum B_{cor} is found by the approximate distribution $B_{cor} = (B'L')^{0.5}$ where $B' = B + z$ and $L' = L + z$, B = foundation width, L = foundation length, and z = depth to top of the compressible soil profile. Substitute B_{cor} for B in Figure 3-15. $\rho_{\lambda c}$ is the corrected consolidation settlement. This correction should not be applied to bonded clays.

(b) The rebound loop in the laboratory curve is needed to develop the recompression line BF. Evaluation of the recompression index C_r is illustrated in Figure 3-13. The recompression index is equal to or slightly smaller than the swelling index, C_s. Approximate correlations of the swelling index are shown in Figure 3-14.

(c) Settlement ρ_{cj} of stratum j in in. may be estimated as the sum of recompression and virgin consolidation settlements. The final void ratio is found graphically from Figure 3-12b. The change in void ratio may be calculated by

$$\Delta e_j = C_r \cdot \log_{10} \frac{\sigma'_{pj}}{\sigma'_{oj}} + C_c \cdot \log_{10} \frac{\sigma'_{fj}}{\sigma'_{pj}} \quad (3\text{-}23)$$

where C_r is the average slope of the recompression line BF. If $\sigma'_{fj} < \sigma'_{pj}$, ignore the right-hand term in Equation

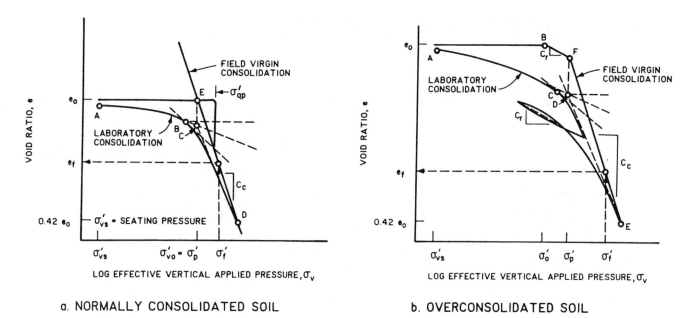

a. NORMALLY CONSOLIDATED SOIL   b. OVERCONSOLIDATED SOIL

Figure 3-12. Construction of field virgin consolidation relationships

Table 3-6. Reconstruction of Virgin Field Consolidation (Data from Item 54).

Step (1)	Description (2)
	a. Normally Consolidated Soil (Figure 3-12a)
1	Plot point B at the point of maximum radius of curvature of the laboratory consolidation curve.
2	Plot point C by the Casagrande construction procedure: (1) Draw a horizontal line from B; (2) draw a line tangent to the laboratory consolidation curve through B; and (3) draw the bisector between horizontal and tangent lines. Point C is the intersection of the straight portion of the laboratory curve with the bisector. Point C indicates the maximum past pressure σ'_p.
3	Plot point E at the intersection e_o and σ'_p. e_o is given as the initial void ratio prior to testing in the consolidometer and σ'_p is found from step 2.
4	Plot point D at the intersection of the laboratory virgin consolidation curve with void ratio $e = 0.42e_o$.
5	The field virgin consolidation curve is the straight line determined by points E and D.
	b. Overconsolidated Soil (Figure 3-12b)
1	Plot point B at the intersection of the given e_o and the initial estimated in situ effective overburden pressure σ'_o.
2	Draw a line through B parallel to the mean slope C_r of the rebound laboratory curve.
3	Plot point D using step 2 in Table 3-6a for normally consolidated soil.
4	Plot point F by extending a vertical line through D up through the intersection of the line of slope C_r extending through B.
5	Plot point E at the intersection of the laboratory virgin consolidation curve with void ratio $e = 0.42e_o$.
6	The field virgin consolidation curve is the straight line through points F and E.

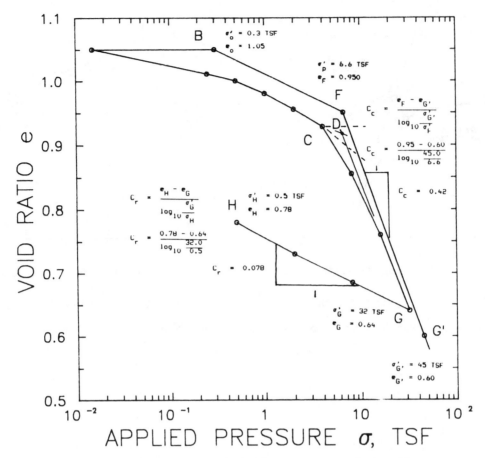

Figure 3-13. Example void ratio–logarithm pressure relationship

3-23 containing C_c and substitute σ'_{fj} for σ'_{pj} in the term containing C_r. Settlement of stratum j is found from Equation 3-20 and ultimate settlement ρ_c of compressible soil in the profile is found from Equation 3-22.

Table 3-7. Estimates of the Virgin Compression Index, C_c

Soil (1)	C_c (2)
Organic soils with sensitivity less than 4	$0.009(LL - 10)$
Organic soils, peat	$0.0115W_n$
Clay	$1.15(e_o - 0.35)$
Clay	$0.012W_n$
Clay	$0.01(LL - 13)$
Varved clays	$(1 + e_o) \cdot [0.1 + 0.006(W_n - 25)]$
Uniform silts	0.20
Uniform sand, loose	0.05 to 0.06
Uniform sand, dense	0.02 to 0.03

Note: LL = liquid limit in percent; W_n = natural water content in percent; and e_o = initial void ratio.

(5) Underconsolidated Soil. Occasionally, a compressible soil stratum may be found to have excess hydrostatic pore pressures such as when the stratum had not reached equilibrium pore water pressures under existing overburden pressures or the groundwater level had been lowered. The effective stress will increase as the pore pressures dissipate and cause recompression settlement until the effective stress equals the preconsolidation stress. Virgin consolidation settlement will continue to occur with increasing effective stress until all excess pore pressures are dissipated. If the initial effective stress is less than the preconsolidation stress σ'_p, then the ultimate settlement may be found as for an overconsolidated soil from Equations 3-20 and 3-23. σ'_{oj} is the initial effective stress found from Equation 1-1, the initial total overburden pressure minus the initial total pore water pressure. σ'_{fj} is the final effective stress found from the final total overburden pressure minus the equilibrium or final pore water pressure. If σ'_{oj} equals σ'_{pj}, then the ultimate settlement may be found as for a normally consolidated soil from Equations 3-20 and 3-21.

B. ADJUSTMENT FOR OVERCONSOLIDATION EFFECTS. The effects of overconsolidation

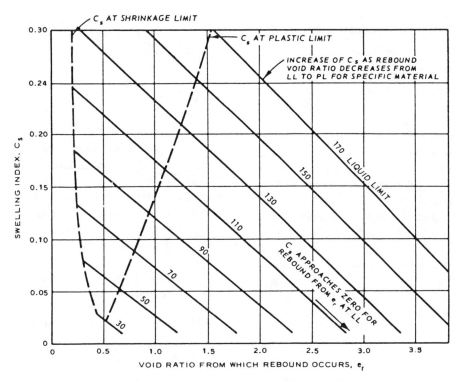

Figure 3-14. Approximate correlations for the swelling index of silts and clays (Figure 3-10, TM 5-818-1)

and departure from 1-D compression on the initial excess pore pressure may require correction to the calculated settlement and rate of settlement. The following semiempirical procedures have been used to correct for these effects. Numerical methods of analysis offer a rational alternative approach to include 3-D effects, but these have not proved useful in practice.

(1) Skempton and Bjerrum Correction. The corrected consolidation settlement $\rho_{\lambda c}$ of a clay stratum is found by

$$\rho_{\lambda c} = \lambda \rho_c \qquad (3\text{-}24)$$

where λ is the settlement correction factor (Figure 3-15). The equivalent dimension of the loaded area should be corrected to the top of the compressible stratum by the approximate stress distribution method as illustrated in step 9 in Table 3-5 or Appendix C. The corrected settlement is still assumed to be 1-D, although overconsolidation effects are considered. $\lambda = 1$ if $B/H > 4$ or if depth to the compressible stratum is $> 2B$.

(2) Stress Path Correction. This alternative approach attempts to simulate stress paths that occur in the field, as illustrated in Table 3-8. This procedure may require special laboratory tests using triaxial cells capable of undrained loading followed by

consolidation. These tests have not usually been performed and are without standard operating guidelines. Approximations necessary to estimate suitable points in the soil profile for testing and estimates of stresses

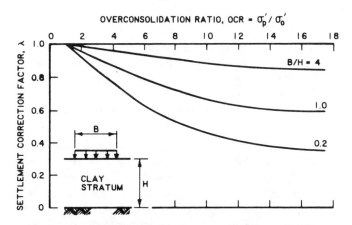

Figure 3-15. Settlement correction factor for overconsolidation effects. Reprinted by permission of the Transportation Research Board, National Research Council from *Special Report 163*, 1976, "Estimating Consolidation Settlement of Shallow Foundations On Overconcolidated Clay," by G. A. Leonards, p. 15

Table 3-8. Summary of the Stress Path Procedure (Data from Item 35)

Step (1)	Procedure (2)
1	Select one or more points within the soil profile beneath the proposed structure.
2	Determine initial stresses and pore pressures at the selected points.
3	Estimate for each point the stress path for loading to be imposed by the structure. The stress path usually depends on undrained loading initially, followed by consolidation.
4	Perform laboratory tests that follow the estimated stress paths; duplicate initial stresses, measure strains from undrained loading, then consolidate to the final effective stress σ'_f and measure strains.
5	Use the strains measured to estimate settlement of the proposed structure.

applied to soil elements at the selected points may introduce errors more significant than the Skempton and Bjerrum correction procedure.

3-13. Time Rate of Settlement

The solution for time rate of primary consolidation settlement is based on the Terzaghi 1-D consolidation theory in which settlement as a function of time is given by

$$\rho_{ct} = \frac{U_t \cdot \rho_{\lambda c}}{100} \qquad (3\text{-}25)$$

where ρ_{ct} = consolidation settlement at time t in ft; U_t = degree of consolidation of the compressible stratum at time t in percent; and $\rho_{\lambda c}$ = ultimate consolidation settlement adjusted for overconsolidation effects in ft. Refer to Table 3-9 for the general procedure to determine time rate of settlement from primary consolidation.

A. EVALUATION OF THE DEGREE OF CONSOLIDATION. Solution of the Terzaghi consolidation theory to determine U_t is provided in Table 3-10 as a function of time factor T_v for four cases of different distributions of the initial excess pore water pressure. Figure 3-16 illustrates example distributions of the initial excess pore water pressure for single drainage (from one surface) and double drainage (from top and bottom surfaces).

(1) Time Factor. The time factor is given by

$$T_v = \frac{C_v t}{H_e^2} \qquad (3\text{-}26)$$

where c_v = coefficient of consolidation of the stratum in ft/day; and H_e = equivalent height of the compressible stratum in ft.

The equivalent thickness of a compressible stratum for single drainage (drainage from one boundary) is the actual height of the stratum. H_e is 1/2 of the actual height of the stratum for double drainage (drainage from top and bottom boundaries).

(2) Coefficient of Consolidation. The coefficient of consolidation c_v may be found experimentally from conventional (step load) laboratory 1-D consolidometer test results by four methods described in Table 3-11, Figure 3-17, and Appendix E. Both Casagrande and Taylor methods (Table 3-11) are recommended, and they may provide reasonable lower- and upper-bound values of the coefficient of consolidation. The Casagrande logarithm time method is usually easier to use with the less pervious cohesive soils, whereas the Taylor square root of time method is easier to use with the more pervious cohesionless soils.

(a) The Casagrande logarithm time method (Figure 3-17a) determines

$$c_v = \frac{0.197 h_e^2}{t_{50}} \qquad (3\text{-}27a)$$

where c_v = coefficient of consolidation of stratum in ft²/day; h_e = equivalent specimen thickness in ft; and t_{50} = time at 50 percent of primary consolidation in days.

The equivalent specimen thickness is the actual specimen height for single drainage and 1/2 of the specimen height for double drainage. This method usually provides a low value or slow rate of consolidation.

(b) The Taylor square root of time method (Figure 3-17b), determines

$$c_v = \frac{0.848 h_e^2}{t_{90}} \qquad (3\text{-}27b)$$

This method usually calculates a faster rate of consolidation than the Casagrande method and may better simulate field conditions.

(c) c_v should be plotted as a function of the applied consolidation pressure. An appropriate value of c_v can be selected based on the final effective pressure σ'_f of the soil for a specific case.

(d) Figure 3-18 illustrates empirical correlations of the coefficient of consolidation with the liquid limit.

(e) The procedure shown in Table 3-12 should be used to transform a compressible soil profile with variable coefficients of consolidation to a stratum of equivalent thickness H' and coefficient of consolidation c_v. T_v may be calculated from Equation 3-26 with $H_e = H'$. Refer to 3-13d, "Internal Drainage Layers," to es-

Table 3-9. Time Rate of Settlement

Step (1)	Description (2)
1	Evaluate lower- and upper-bound values of the coefficient of consolidation, c_v, of each soil stratum in the profile for each consolidation load increment from deformation-time plots of data from 1-D consolidometer tests. Plot c_v as a function of the logarithm of applied pressure. Refer to Table 3-11 and Figure 3-17 for methods of calculating c_v.
2	Select appropriate values of c_v from the c_v versus logarithm pressure plots using σ_f' found from step 5 in Table 3-5. Preliminary estimates of c_v may be made from Figure 3-18.
3	Select minimum and maximum values of c_v and calculate the effective thickness H' of a multilayer soil profile using the procedures in Table 3-12 relative to one of the soil layers with a given c_{vi}. If the soil profile includes pervious incompressible seams, then evaluate T_v and U_t in steps 4 to 6 for each compressible layer and calculate U_t of the soil profile by step 7.
4	Evaluate minimum and maximum time factors T_v of the compressible soil profile from Equation 3-26 $$T_v = \frac{c_v t}{H_e^2}$$ for various times t using c_v from step 3 (or c_{vi} for multilayer soil). The equivalent compressible soil height H_e is 1/2 of the actual height (or 1/2 of the effective height H' of multiple soil layers) for double drainage from top and bottom surfaces of the compressible soil and equal to the height of the compressible soil for single drainage.
5	Select the case (Table 3-10 and Figure 3-16) that best represents the initial pore water pressure distribution. If none of the given pressure distributions fit the initial distribution, then approximate the initial distribution as the sum or difference of some combination of the given standard distributions in Table 3-10 as illustrated in Figure 3-19. Note the cases and relative areas of the standard pore water pressure distributions used to approximate the initial distribution.
6	Evaluate minimum and maximum values of the degree of consolidation U_t for given T_v from Table 3-10. If none of the four cases in Table 3-10 model the initial pore pressure distribution, then the overall degree of consolidation may be evaluated by dividing the pore pressure distribution into areas that may be simulated by the cases in Table 3-10 and using Equation 3-28 $$U_t = \frac{U_{t1}A_1 + U_{t2}A_2 + \cdots U_{ti}A_i}{A}$$ where U_{ti} = degree of consolidation of case i, i = 1 to 4; A_i = area of pore pressure distribution of case i; and A = area of approximated pore pressure distribution. U_t may also be the degree of consolidation of a soil bounded by internal drainage layers (pervious soil). Omit step 7 if U_t is the degree of consolidation of the soil where pervious seams are not present.
7	Evaluate the influence of internal drainage layers (pervious seams) on settlement by (Equation 3-29) $$U_t = \frac{1}{\rho_c} \cdot (U_{t1}\rho_{c1} + U_{t2}\rho_{c2} + \cdots U_{tn}\rho_{cn})$$ where U_t is the degree of consolidation at time t and ρ_c is the ultimate consolidation settlement of the compressible soil profile. Subscripts 1, 2, . . . , n indicate each compressible layer between seams.
8	Determine the consolidation settlement as a function of time ρ_{ct}, where $\rho_{ct} = U_t \cdot \rho_c$ (Equation 3-25).

Table 3-10. Degree of Consolidation as a Function of Time Factor T_v

| T_v (1) | Average Degree of Consolidation, U_t (percent) | | | |
	Case 1[a] (2)	Case 2 (3)	Case 3 (4)	Case 4 (5)
0.004	7.14	6.49	0.98	0.80
0.008	10.09	8.62	1.95	1.60
0.012	12.36	10.49	2.92	2.40
0.020	15.96	13.67	4.81	4.00
0.028	18.88	16.38	6.67	5.60
0.036	21.40	18.76	8.50	7.20
0.048	24.72	21.96	11.17	9.69
0.060	27.64	24.81	13.76	11.99
0.072	30.28	27.43	16.28	14.36
0.083	32.51	29.67	18.52	16.51
0.100	35.68	32.88	21.87	19.77
0.125	39.89	36.54	26.54	24.42
0.150	43.70	41.12	30.93	28.86
0.175	47.18	44.73	35.07	33.06
0.200	50.41	48.09	38.95	37.04
0.250	56.22	54.17	46.03	44.32
0.300	61.32	59.50	52.30	50.78
0.350	65.82	64.21	57.83	56.49
0.400	69.79	68.36	62.73	61.54
0.500	76.40	76.28	70.88	69.95
0.600	81.56	80.69	77.25	76.52
0.800	88.74	88.21	86.11	85.66
1.000	93.13	92.80	91.52	91.25
1.500	98.00	97.90	97.53	97.45
2.000	99.42	99.39	99.28	99.26

[a] See Figure 3-16.

timate U_t of a soil profile with pervious incompressible sand seams interspersed between compressible soil.

B. SUPERPOSITION OF EXCESS PORE PRESSURE DISTRIBUTION. An initial pore pressure distribution that is not modeled by any of the four cases in Table 3-10 and Figure 3-16 may sometimes be approximated by superposition of any of the four cases and the overall or weighted degree of consolidation found by

$$U_t = \frac{U_{t1}A_1 + U_{t2}A_2 + \cdots U_{ti}A_i}{A} \quad (3-28)$$

where A represents the areas of the initial pore pressure distributions. Subscripts $1, 2, \ldots, i$ indicate each pore pressure distribution. Linearity of the differential equations describing consolidation permits this assumption.

(1) Example of Excess Pore Water Pressure Distributions. Some examples of complex excess pore water pressure distributions are shown in Figure 3-19.

(2) Application. For single drainage, a decreasing excess pore pressure distribution may be modeled as shown in Figure 3-19b. If $T_v = 0.2$, the degree of consolidation is 50.41 and 37.04 percent for cases 1 and 4, respectively (Table 3-10). The overall degree of consolidation from Equation 3-27 for the example in Figure 3-19b is

$$U_t = \frac{50.41 \cdot H_e \cdot H_o - 37.04 \cdot 0.5 \cdot H_e \cdot H_o/2}{H_e \cdot H_o - 0.5 \cdot H_e \cdot H_o/2}$$

$$U_t = \frac{50.41 \cdot 1.0 - 37.04 \cdot 0.25}{1.0 - 0.25} = 54.87 \text{ percent}$$

The total area of the complex pore pressure distribution equals the area of case 1 less the area of case 4 (Figure 3-19b).

C. INTERNAL DRAINAGE LAYERS. Internal drainage layers of pervious soil within an otherwise low-permeable clay stratum will influence the rate of settlement. This influence can be considered by summation of the degrees of consolidation of each compressible layer between the pervious seams by (item 52)

$$U_t = \frac{1}{\rho_c} \cdot (U_{t1} \cdot \rho_{c1} + U_{t2} \cdot \rho_{c2} + \cdots + U_{tn} \cdot \rho_{cn}) \quad (3-29)$$

where U_t is the degree of consolidation at time t and ρ_c is the ultimate consolidation settlement of the entire compressible stratum. The subscripts $1, 2, \ldots, n$ indicate each compressible layer between pervious seams.

D. TIME-DEPENDENT LOADING. The rate of load application to foundation soils is usually time-dependent. Estimates of the degree of consolidation of time-dependent loads may be made by dividing the total load into several equal and convenient increments such as the 25 percent increments illustrated in Figure 3-20. Each increment is assumed to be placed instantaneously at a time equal to the average of the starting and completion times of the placement of the load increment. The degree of consolidation U of the underlying compressible soil is evaluated for each of the equal load increments as a function of time and divided by the number of load increments to obtain a weighted U. Only one curve need be evaluated for the soil if the thickness of the compressible stratum and coefficient of consolidation are constant. The weighted U of each load increment may then be summed graphically as illustrated in Figure 3-20 to determine the de-

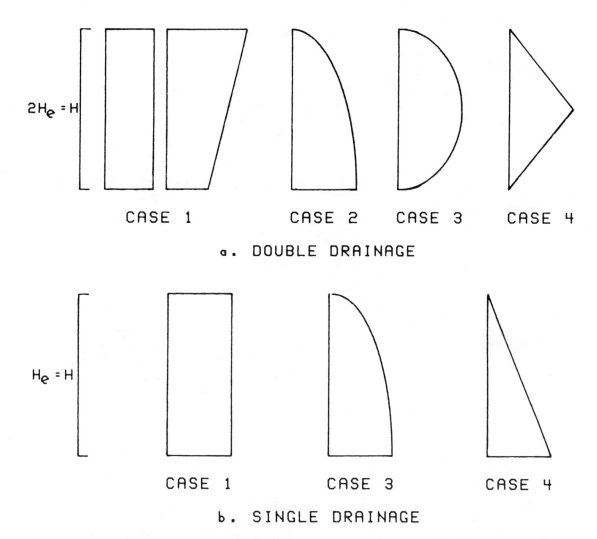

Figure 3-16. Example distributions of excess pore water pressure for double and single drainage; *H* is the actual stratum thickness and *H$_e$* is the equivalent height

Table 3-11. Evaluation of the Coefficient of Consolidation *c$_v$* by the Step Load Test (Data from Item 66)

Method (1)	Equation for *c$_v$* (2)	Procedure (3)	Example (4)
Terzaghi	$\dfrac{0.848h_e^2}{t_{90}}$	1. Measure initial specimen height h_o and set initial dial reading d_o. 2. Measure dial reading *d* as a function of time *t* and final specimen height h_f. Plot *d* versus $\log_{10} t$. Determine d_s, the corrected zero point ($d_o - d_s$ = initial compression), by measuring vertical distance between time of about 0.1 min and a time that is 4 times this. 3. Determine time t_{100} and compression d_{100} to 100 percent consolidation as	$h_o = 1.148$ in. $d_o = 0.0000$ in. $h_f = 1.140$ in. $d_s = 0.0002$ in. Refer to Figure 3-17a $t_{100} = 392$ min or 0.27 day $d_{100} = 0.0118$ in.

Table 3-11. Continued

Method (1)	Equation for c_v (2)	Procedure (3)	Example (4)
		the intersection of the tangent and asymptote of the consolidation curve. Then determine d_{90} and t_{90}.	$d_{00} = d_s + 0.9(d_{100} - d_s)$ $= 0.0002 +$ $0.9(0.0118 -$ $0.0002)$ $= 0.0002 + 0.0104$ $= 0.0106$ in. $t_{90} = 290$ min or 0.20 day from Figure 3-17a
		4. Determine equivalent thickness of drainage path, $h_e = (h_o + h_f)/2$ if single drainage; $h_e = (h_o + h_f)/4$ if double drainage. 5. Calculate c_v.	$h_e = (h_o + h_f)/4$ $= 2.288/4 = 0.572$ in. or 0.0477 ft $c_v = \dfrac{0.85 \cdot 0.0477^2}{0.20} =$ 0.010 ft^2/day
Casagrande	$\dfrac{0.197h_e^2}{t_{50}}$	Same as Terzaghi except determine time to reach 50 percent of consolidation t_{50}.	$d_{50} = d_s + 0.5(d_{100} - d_s)$ $= 0.0002 +$ $0.5(0.0118 -$ $0.0002)$ $= 0.0002 + 0.0058$ $= 0.0060$ in. $t_{50} = 90$ min or 0.063 day $c_v = \dfrac{0.197 \cdot 0.0477}{0.063} =$ 0.007 ft^2/day
Inflection point	$\dfrac{0.450h_e^2}{t_i}$	Same except note time to reach inflection point of curve t_i.	$t_i = 200$ min or 0.14 day from Figure 3-17a $c_v = \dfrac{0.405 \cdot 0.0477^2}{0.14} =$ 0.007 ft^2/day
Taylor	$\dfrac{0.848h_e^2}{t_{90}}$	1. Measure initial specimen height h_o and set initial dial reading d_o. 2. Measure dial reading d as a function of time t and final specimen height h_f. Plot d versus $\sqrt{t}$ or square root of time. 3. Extend straight line portion back to $\sqrt{t} = 0$ to obtain corrected initial reading d_o. 4. Through d_o draw a straight line with inverse slope 1.15 times tangent and intersect laboratory curve to obtain t_{90}. 5. Determine h_e as above. 6. Calculate c_v.	$h_o = 1.148$ in. $d_o = 0.0000$ in. $h_f = 1.140$ Refer to Figure 3-17b $d_o = 0.0005$ in. from Figure 3-17b $\sqrt{t_{90}} = 18.4$ or $t_{90} = 339$ min or 0.23 day Refer to Figure 3-17b $h_e = 0.0477$ ft $c_v = \dfrac{0.848 \cdot 0.0477^2}{0.23} =$ 0.008 ft^2/day

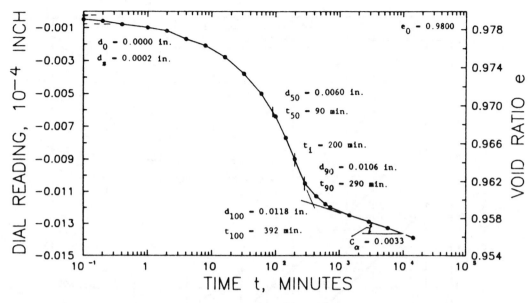

a. VOID RATIO-LOGARITHM TIME

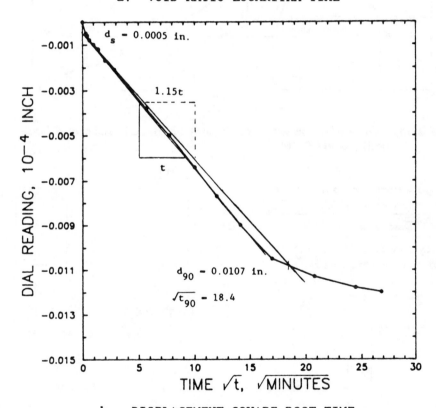

b. DISPLACEMENT-SQUARE ROOT TIME

Figure 3-17. Example time plots from one-dimensional consolidometer test, $\Delta\sigma = 1$ TSF

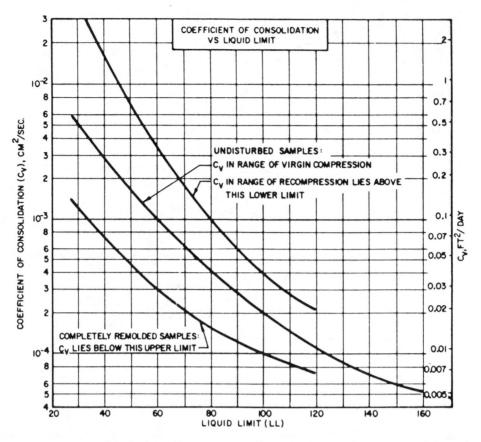

Figure 3-18. Correlations between coefficient of consolidation and liquid limit (NAVFAC DM 7.1)

Table 3-12. Procedure to Evaluate the Effective Thickness and Average Degree of Consolidation for Multiple Soil Layers (After NAVFAC DM 7.1)

Step (1)	Description (2)
1	Select any layer i with coefficient of consolidation c_{vi} and thickness H_i.
2	Transform the thickness of every other layer to an effective thickness H_i' $$H_1' = H_1 \left[\frac{c_{vi}}{c_{v1}}\right]^{1/2}$$ $$H_2' = H_2 \left[\frac{c_{vi}}{c_{v2}}\right]^{1/2}$$ $$H_n' = H_n \left[\frac{c_{vi}}{c_{vn}}\right]^{1/2}$$
3	Calculate the total effective thickness by $$H' = H_1' + H_2' + \cdots + H_i' + \cdots + H_n'$$
4	Treat the entire thickness as a single layer of effective thickness H' with a coefficient of consolidation $c_v = c_{vi}$ and evaluate the time factor T_v from Equation 3-26. Evaluate the degree of consolidation with the assistance of Table 3-10 and Figure 3-16.

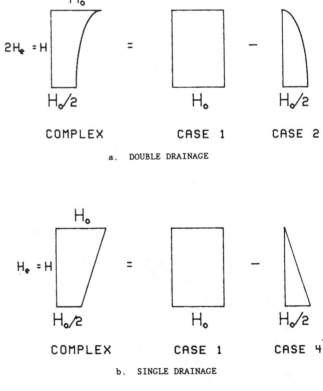

Figure 3-19. Example complex excess pore water pressure distributions

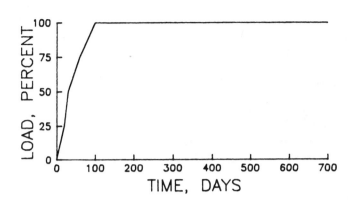

a. TIME-DEPENDENT LOADING

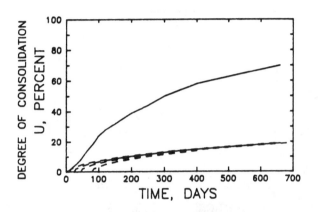

b. DEGREE OF CONSOLIDATION

Figure 3-20. Degree of consolidation for time-dependent loading

gree of consolidation of the time-dependent loading. Chapter 5 of NAVFAC DM 7.1 provides a nomograph for evaluating U for a uniform rate of load application.

3-14. Example Application of Primary Consolidation

An embankment (Figure 3-21) is to be constructed on a compressible clay stratum 20 ft thick. The groundwater level is at the top of the compressible clay stratum. A consolidometer test was performed on an undisturbed specimen of the soil stratum after the standard load procedure described in EM 1110-2-1906. The specimen was taken from a depth of 10 ft and drainage was allowed on both top and bottom surfaces. A plot of the laboratory consolidation void ratio versus logarithm pressure relationship is shown in Figure 3-13.

A. ULTIMATE PRIMARY CONSOLIDATION. The procedure described in Table 3-5 was applied to evaluate ultimate settlement beneath the

edge and center of the embankment by hand calculations. The solution is worked out in Table 3-13a.

B. TIME RATE OF CONSOLIDATION. The procedure described in Table 3-9 was applied to evaluate the rate of settlement beneath the edge and center of the embankment by hand calculations assuming an instantaneous rate of loading. The solution is worked out in Table 3-13b.

3-15. Accuracy of Settlement Predictions

Experience shows that predictions of settlement are reasonable and within 50 percent of actual settlements for many soil types. Time rates of settlement based on laboratory tests and empirical correlations may not be representative of the field because time rates are influenced by in situ fissures, existence of high permeable sand or low permeable bentonite seams, impervious boundaries, and nonuniform soil parameters as well as the rate of construction.

A. PRECONSOLIDATION STRESS. Soil disturbance of laboratory samples used for 1-D consolidation tests decreases the apparent preconsolidation stress.

B. VIRGIN COMPRESSION INDEX. Soil disturbance decreases the compression index.

C. SWELLING AND RECOMPRESSION INDICES. Soil disturbance increases the swelling and recompression indices.

D. COEFFICIENT OF CONSOLIDATION. Soil disturbance decreases the coefficient of consolidation for both virgin compression and recompression (Figure 3-18) in the vicinity of initial overburden and preconsolidation stresses. The value of c_v decreases abruptly at the preconsolidation stress for good undisturbed samples.

E. FIELD TEST EMBANKMENT. A field test embankment may be constructed for significant projects to estimate field values of soil parameters such as C_c and c_v. Installation of elevation markers, inclinometers, and piezometers allow the measurement of settlement, lateral movement, and pore pressures as a function of time. These field soil parameters may subsequently be applied to full-scale structures.

3-16. Computer Solutions

Several computer programs are available to expedite calculation of settlement and rates of settlement of structures constructed on multilayer soil profiles.

A. VERTICAL STRESS DISTRIBUTION. The vertical stress distribution from the Boussinesq and

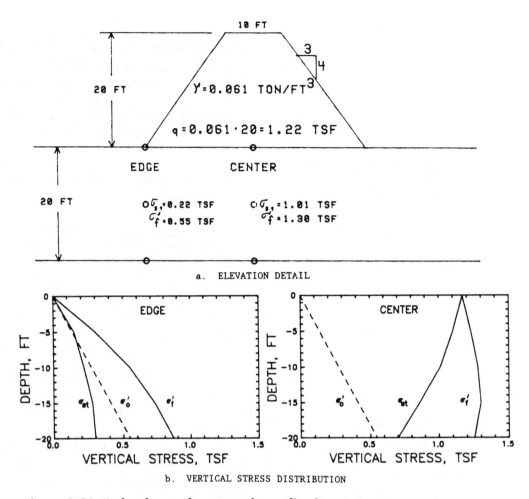

Figure 3-21. Embankment for example application

Table 3-13. Evaluation of Consolidation Settlement by Hand Calculations for Example Application of Embankment (Figure 3-21)

Step (1)	Description (2)
	a. Total Settlment
1	The preconsolidation stress σ'_p after the Casagrande construction procedure (Table 3-6a) is 6.6 tsf shown in Figure 3-13 (neglecting a minimum and maximum range). Since $\sigma'_p > \sigma'_o$, the soil is overconsolidated with an OCR of about 22. The field virgin consolidation line is evaluated by the procedure in Table 3-6b.
2	The initial effective stress distribution σ'_o is shown in Figure 3-21b. The wet unit weight γ is 0.061 ton/ft^3 and γ_w is 0.031 ton/ft. σ'_o at 10 ft of depth is 0.3 tsf.
3	The initial void ratio of the specimen prior to consolidation is e_o = 1.05 (Figure 3-13).
4	The virgin compression index C_c = 0.42 and the recompression index C_r = 0.078 (Figure 3-13).
5	The pressure distribution applied by the embankment was calculated using the trapezoidal distribution [Table C-1b (Appendix C)]. At 10 ft below ground surface the vertical stress applied by the embankment at the edge is 0.22 tsf and at the center it is 1.01 tsf. The final effective pressure σ'_f 10 ft below ground surface at the edge is 0.55 tsf, at the center 1.30 tsf. The estimated pressure distributions are shown in Figure 3-21b.
6	The change in void ratio Δe at the 10-ft depth may be found from Equation 3-23 where the right-hand part of the equation containing C_c is ignored because $\sigma'_f < \sigma'_p$ $$\Delta e = C_r \cdot \log \frac{\sigma'_f}{\sigma'_o} \quad \text{Edge:} \quad \Delta e_e = 0.078 \cdot \log_{10} \frac{0.55}{0.30} = 0.020$$ $$\text{Center: } \Delta e_c = 0.078 \cdot \log_{10} \frac{1.30}{0.30} = 0.050$$

Table 3-13. Continued

Step (1)	Description (2)
	a. Total Settlment

7, 8	Settlement of the stratum from Equation 3-20 is

$$\text{Edge:} \quad \rho_{ce} = \frac{0.020}{1.00 + 1.05} \cdot 20 = 0.195 \text{ ft or } 2.34 \text{ in.}$$

$$\text{Center:} \quad \rho_{cc} = \frac{0.050}{1.00 + 1.05} \cdot 20 = 0.488 \text{ ft or } 5.85 \text{ in.}$$

Improved reliability may be obtained by testing additional specimens at different depths within the compressible stratum, calculating settlements within smaller depth increments and adding the calculated settlements (Equation 3-22).

9　Settlement may be corrected for overconsolidation effects after the Skempton and Bjerrum procedure (Equation 3-24). λ is about 0.8 from Figure 3-15 for an overconsolidation ratio > 18.

$$\text{Edge:} \quad \rho_{3ce} = 0.8 \cdot 2.34 = 1.87 \text{ in.}$$

$$\text{Center:} \quad \rho_{3cc} = 0.8 \cdot 5.85 = 4.68 \text{ in.}$$

	b. Time Rate of Settlement

1, 2	Minimum and maximum estimates of the coefficient of consolidation may be made using the methods in Table 3-11 from a plot of the deformation as a function of time data (Figure 3-17). These data indicate c_v values from 0.007 ft^2/day to 0.010 ft^2/day (Table 3-11). The range of applied consolidation pressures is from 1 to 2 tsf using double drainage during the consolidation test.
3	The time factors for the range of c_v from 0.007 to 0.010 ft^2/day is (Equation 3-26)

$$\frac{0.007t}{10^2} < T_v < \frac{0.010t}{10^2} = 0.00007t < T_v < 0.000010t$$

where the time t is in days. The compressible stratum is assumed to drain on both top and bottom surfaces; therefore, the equivalent height H_e is 10 ft.

4　The excess pore water pressure distribution given by σ_{st} in Figure 3-21 appears to be similar to case 2 at the edge and case 1 at the center (Figure 3-16a). The average degree of consolidation in percent after 1, 10, and 50 years using Table 3-10 is

Time		T_v		U_t (percent)			
Years	Days	Min	Max	Edge (Case 2)		Center (Case 1)	
				Min	Max	Min	Max
1	364	0.025	0.036	14.7	18.8	17.1	21.4
10	3,640	0.255	0.364	54.7	65.4	56.7	67.0
50	18,200	1.274	3.640	95.6	98.9	95.8	98.9

5	Cases 1 and 2 of Figure 3–16a are considered representative of the initial excess pore water pressure distributions so that superposition of the cases in Table 3-9, steps 5 and 6, is not necessary.
6	Consolidation settlement as a function of time t is, Equation 3-24

	Settlement ρ_{ct} (in.)			
	Edge		Center	
Time (years)	Min	Max	Min	Max
1	0.27	0.35	0.80	1.00
10	1.02	1.22	2.65	3.14
50	1.79	1.85	4.48	4.63
∞	1.87		4.68	

Westergaard solutions may be computed beneath foundations, single and multiple footings, and embankments from Program CSETT (item 61) and Program IOO16 (item 45).

B. ULTIMATE CONSOLIDATION SETTLEMENT. Long-term consolidation settlement of structures may be computed assuming the Terzaghi 1-D consolidation by Program MAGSETTI (item 45) using output from Program CSETT or IOO16.

C. ULTIMATE CONSOLIDATION AND RATE OF SETTLEMENT. One-dimensional consolidation settlement and rates of settlement by Terzaghi 1-D consolidation theory may be computed by Program FD31 (item 45) and Program CSETT (item 61).

(1) Program FD31 does not consider the influence of the vertical stress distribution with depth, and therefore it is applicable to fills or embankments with lateral dimensions substantially greater than the thickness of the consolidating soil.

(2) Program CSETT considers loaded regions of simple and complex geometric shapes for single or multiple and time-dependent loads. Loads may be 2- or 3-D. Stress distributions may be calculated by either Boussinesq or Westergaard methods. The program allows analysis of multiple soil layers and a variety of drainage conditions. Output consists of total settlement, settlement of individual layers, and degree of consolidation as a function of time and location requested by the user.

D. SETTLEMENT OF SOFT SOIL. Settlement from desiccation and consolidation in soft, compressible soil with large void ratios may be computed by program PCDDF (item 8). This program is applicable to dredged material and considers time-dependent loads, influence of void ratio on self-weight, permeability, nonlinear effective stress relationships, and large strains.

E. SETTLEMENT OF SHALLOW FOUNDATIONS IN SAND. Corps Program IOO30, CSANDSET, can calculate the immediate settlement in sands of 14 different procedures including Alpan, Schultze and Sherif, Terzaghi and Peck, Schmertmann, and elastic methods. Program IOO30 considers water table depth, embedment depth, and foundation dimensions for a variety of soil conditions in multilayer sands. Soil input data include SPT, CPT, elastic modulus, and water table depth.

F. VERTICAL DISPLACEMENT OF VARIOUS SOIL TYPES. Appendix F provides a user's manual and listing of computer program VDISPL for calculating immediate settlement of granular materials using Schmertmann's procedure modified to consider prestress. Program VDISPL can also calculate immediate settlement of an elastic soil, consolidation/swell of an expansive soil, and settlement of a collapsible soil

(see Chapter 5). Finite element program CON2D (item 15) may be used to calculate plane strain 2-D consolidation settlement of embankments and structures on multiple soil layers using the Cam Clay elasto-plastic constitutive soil model. CON2D may also analyze consolidation of saturated and partly saturated earth masses for time-dependent vertical loads to determine settlement, rate of settlement, and pore pressure distribution. This program may analyze the condition of saturated and partly saturated earth mass.

Section IV. Secondary Compression and Creep

3-17. Description

Secondary compression and creep are time-dependent deformations that appear to occur at essentially constant effective stress with negligible change in pore water pressure. Secondary compression and creep may be a dispersion process in the soil structure causing particle movement, and they may be associated with electrochemical reactions and flocculation. Although creep is caused by the same mechanism as secondary compression, they differ in the geometry of confinement. Creep is associated with deformation without volume and pore water pressure changes in soil subject to shear, whereas secondary compression is associated with volume reduction without significant pore water pressure changes.

A. MODEL. Secondary compression and creep may be modeled by empirical or semiempirical viscoelastic processes in which hardening (strengthening) or softening (weakening) of the soil occurs. Hardening is dominant at low stress levels, whereas weakening is dominant at high stress levels. Deformation in soil subject to a constant applied stress can be understood to consist of three stages. The first stage is characterized by a change in the rate of deformation that decreases to zero. The second or steady-state stage occurs at a constant rate of deformation. A third stage may also occur at sufficiently large loads in which the rate of deformation increases, ending in failure as a result of weakening in the soil. Soil subject to secondary compression in which the volume decreases, as during a 1-D consolidometer test, may gain strength or harden with time leading to deformation that eventually ceases, and therefore the second (steady-state) and third (failure state) stages may never occur.

B. RELATIVE INFLUENCE. Secondary compression and creep are minor relative to settlement caused by elastic deformation and primary consolidation in many practical applications. Secondary compression may contribute significantly to settlement where soft soil exists, particularly soft clay, silt, and soil containing organic matter such as peat or Muskeg or where a deep compressible stratum is subject to small pressure increments relative to the magnitiude of the effective consolidation pressure.

3-18. Calculation of Secondary Compression

Settlement from secondary compression ρ_s has been observed from many laboratory and field measurements to be approximately a straight line on a semi-logarithmic plot with time (Figure 3-17a) following completion of primary consolidation. The decrease in void ratio from secondary compression is

$$\Delta e_{st} = C_\alpha \cdot \log \frac{t}{t_{100}} \qquad (3\text{-}30)$$

where Δe_{st} = change in void ratio from secondary compression at time t; C_α = coefficient of secondary compression; t = time at which secondary compression settlement is to be calculated in days; and t_{100} = time corresponding to 100 percent of primary consolidation in days. Secondary compression settlement is calculated from Equation 3-20 in a manner similar to primary consolidation settlement.

A. COEFFICIENT OF SECONDARY COMPRESSION. C_α is the slope of the void ratio-logarithm time plot for time exceeding that required for 100 percent of primary consolidation, t_{100}. t_{100} is arbitrarily determined as the intersection of the tangent to the curve at the point of inflection with the tangent to the straight line portion representing a secondary time effect (Figure 3-17a).

B. ESTIMATION OF C_α. A unique value of C_α/C_c has been observed (Table 3-14) for a variety of different types of soils. The ratio C_α/C_c is constant and the range varies between 0.025 and 0.100 for all soils. High values of C_α/C_c relate to organic soils. C_α will in general increase with time if the effective consolidation pressure σ' is less than a critical pressure or

Table 3-14. Coefficient of Secondary Compression C_α (Data from Item 43)

Soil (1)	C_α/C_c (2)
Clay	0.025 to 0.085
Silt	0.030 to 0.075
Peat	0.030 to 0.085
Muskeg	0.090 to 0.100
Inorganic	0.025 to 0.060

the preconsolidation stress σ'_p. For σ' greater than σ'_p, C_α will decrease with time; however, C_α will remain constant with time within the range of effective pressure $\sigma' > \sigma'_p$ if C_c also remains constant (e.g., the slope of the e-log σ curve is constant for $\sigma' > \sigma'_p$). A first approximation of the secondary compression index C_α is $0.0001 W_n$ for $10 < W_n < 3000$ where W_n is the natural water content in percent (after NAVFAC DM 7.1).

C. ACCURACY. Soil disturbance decreases the coefficient of secondary compression in the range of virgin compression. Evaluation of settlement caused by secondary compression has often not been reliable.

D. EXAMPLE PROBLEM. The coefficient of secondary compression was determined to be 0.0033 and time t_{100} is 392 min or 0.27 day (Figure 3-17a) for this example problem. The change in void ratio after time $t = 10$ years or 3640 days is

$$\Delta e = C_\alpha \cdot \log_{10} \frac{t}{t_{100}} = 0.0033 \cdot \log_{10} \frac{3640}{0.27} = 0.0136$$

$$(3\text{-}31)$$

The settlement from Equation 3-20 for an initial void ratio $e_{100} = 0.96$ is

$$\rho_s = \frac{0.0136}{1 + 0.96} \cdot 20 = 0.139 \text{ ft or } 1.67 \text{ in.}$$

for a stratum of 20 ft thickness.

CHAPTER 4

EVALUATION OF SETTLEMENT FOR DYNAMIC AND TRANSIENT LOADS

4-1. General

Dynamic and transient forces cause particle rearrangements and can cause considerable settlement, particularly in cohesionless soils, when the particles move into more compact positions. A large portion of dynamic live forces applied to foundation soil is from traffic on pavements. Dynamic forces from a rolling wheel depressing a pavement cause a multidirectional combination of cyclic shear and compression strains that precludes presentation of an appropriate settlement analysis in this chapter. This chapter provides guidance for analysis of settlement from earthquakes and repeated loads.

A. AMOUNT OF SETTLEMENT. The amount of settlement depends on the initial density of the soil, thickness of the soil stratum, and maximum shear strain developed in the soil. Cohesionless soils with relative densities D_r greater than about 75 percent should not develop significant settlement; however, intense dynamic loading can cause some settlement of 1 to 2 percent of the stratum thickness even in dense sands.

B. CAUSE OF DIFFERENTIAL SETTLEMENT. A major cause of differential settlement is the compaction of loose sands during dynamic loading. Vibrations caused by machinery often cause differential settlement that may require remedial repairs or limitations on machine operations.

C. TIME EFFECTS. Time required for settlement from shaking can vary from immediately to almost a day. Settlement in dry sands occurs immediately during shaking under constant effective vertical stress. Shaking of saturated sands induces excess pore water pressures, which lead to settlement when the pore pressures dissipate.

D. ACCURACY. Errors associated with settlement predictions from dynamic loads will exceed those for static loads and can be 50 percent or more. These first-order approximations should be checked with available experience.

E. MINIMIZING SETTLEMENT. Dynamic settlement may be insignificant provided that the sum of dynamic and static bearing stresses remains less than 1/2 of the allowable bearing capacity. Settlements that might occur under sustained dynamic loadings may be minimized by precompaction of the soil using dynamic methods. Dynamic compaction subjects the soil to severe dynamic loads that reduce the influence of any later shaking on settlement. Refer to Chapter 6 for dynamic compaction methods of minimizing settlement. Refer to ER 1110-2-1806 for general guidance and direction for seismic design and evaluation for all Corps of Engineer civil works projects.

4-2. Settlement from Earthquakes

Earthquakes primarily cause shear stress, shear strain, and shear motion that propagates toward the ground surface from deep within the earth. This shear can cause settlement initially in deep soil layers followed by settlement in more shallow layers. Settlement caused by ground shaking during earthquakes is often nonuniformly distributed and can cause differential movement in structures leading to major damage. Settlement can occur from compaction in moist or dry cohesionless soil and from dissipation of excess hydrostatic pore pressure induced in saturated soil by earth quake ground motions. Ground motions are multidirectional; however, measurements are generally made in two horizontal and one vertical acceleration components that propagate upward from underlying rock. The vertical component of acceleration is often considered to account for less than 25 percent of the settlement, but this percentage may be exceeded. Soil affected by ground motion and subsequent settlement may extend to considerable depth depending on the source of motion.

A. TENTATIVE SIMPLIFIED PROCEDURE FOR SAND. A tentative simplified procedure to estimate settlement from the shaking forces of earthquakes on saturated sands that are at initial liquefaction and on dry sands is given in Table 4-1. Input data for this procedure include the blowcount N from SPT data as a function of depth, effective and total overburden pressures σ_o' and σ_o, and an estimate of the maximum horizontal acceleration of the ground sur-

Table 4-1. A Suggested Tentative Procedure for Computation of Earthquake Settlement in Sand
(Data from Item 63)

Step (1)	Description (2)
	a. Saturated Sand That Reaches Initial Liquefaction
1	Determine the blowcount N from SPT tests as a function of depth and divide the profile into discrete layers of sand with each layer containing sand with a similar blowcount.
2	Determine correction factor C_{ER} as follows where C_{ER} = estimated rod energy in percent/60.

Location	Hammer	Hammer release	C_{ER}
Japan	Donut	Free-fall	1.3
Japan	Donut	Rope and pulley with special throw release	1.12[a]
United States	Safety	Rope and pulley	1.00[a]
United States	Donut	Rope and pulley	0.75
Europe	Donut	Free-fall	1.00[a]
China	Donut	Free-fall	1.00[a]
China	Donut	Rope and pulley	0.83

[a]Prevalent method at present in United States.

3	Estimate the total and effective overburden pressure σ_o annd σ_o' in tsf from known or estimated soil unit weights and pore water pressures of each layer.
4	Estimate the relative density D_r in percent from results of SPT data using Figure 4-1, improved correlations for overconsolidated soil (item 50), or the expression

$$D_r = 21 \cdot \frac{N_J}{\sigma_o' + 0.7} \qquad (4\text{-}1a)$$

where N_J is the blowcount by Japanese standards and σ_o' is the effective overburden pressure. D_r for normally consolidated material may be estimated by (item 42)

$$D_r = 11.7 + 0.76 \cdot [222N + 1600 - 736\sigma_o' - 50c_u^2]^{1/2} \qquad (4\text{-}1b)$$

where σ_o' = effective overburden pressure in tsf; c_u = uniformity coefficient, D_{60}/D_{10}; D_{60} = grain diameter at which 60 percent of soil weight is finer; and D_{10} = grain diameter at which 10 percent of soil weight is finer.

5	Determine the correction factor C_N from Figure 4-2 using σ_o' and D_r.
6	Calculate normalized blowcount

$$(N_1)_{60} = C_{ER} \cdot C_N \cdot N \qquad (4\text{-}2)$$

where $(N_1)_{60}$ = SPT blowcount normalized to an effective energy delivered to the drill rod at 60 percent of theoretical free-fall energy.

7	Calculate the cyclic shear stress ratio causing initial liquefaction to occur for the given earthquake of magnitude M

$$\left(\frac{\tau_{av}}{\sigma_o'}\right)_M = 0.65 \cdot \frac{a_{max}}{g} \cdot \frac{\sigma_o}{\sigma_o'} \cdot r_d \qquad (4\text{-}3)$$

where τ_{av} = average cyclic shear stress induced by earthquake shaking in tsf; σ_o' = effective overburden pressure in tsf; σ_o = total overburden pressure in tsf; a_{max} = maximum horizontal acceleration of the ground surface in units of g from earthquake records of magnitude M (Regulation Guide 1.60, Nuclear Regulatory Commision; refer to Earthquake Engineering and Geosciences Division, Geotechnical Laboratory, USAE Waterways Experiment Station for Corps of Engineers); g = acceleration of gravity, 32.2 ft/sec/sec; and r_d = stress reduction factor, 1.0 at the ground surface decreasing to 0.9 at depth 30 ft below ground surface.

8	Convert $(\tau_{av}/\sigma_o')_M$ to an equivalent earthquake of magnitude $M = 7.5$ by

$$\left[\frac{\tau_{av}}{\sigma_o'}\right]_{7.5} = \left[\frac{\tau_{av}}{\sigma_o'}\right]_M \cdot \frac{1}{r_m} \qquad (4\text{-}4)$$

Table 4-1. Continued

Step (1)	Description (2)

a. Saturated Sand That Reaches Initial Liquefaction

	where the scaling factor r_m is

Magnitude of earthquake M	No. representative cycles at $0.65\tau_{max}$	r_m
8.5	26	0.89
7.5	15	1.00
6.75	10	1.13
6.00	5	1.32
5.25	2 to 3	1.50

* Representative of the number of equivalent stress cycles caused by the earthquake where τ_{max} = maximum cycle stress

Step	Description
9	Evaluate volumetric strain ϵ_c in percent after initial liquefaction from Figure 4-3 using calculated values of $(N_1)_{60}$ of step 6 and $(\tau_{av}/\sigma_o')_{7.5}$ of step 8.
10	Evaluate earthquake settlement ρ_e after initial liquefaction in inches from $$\rho_e = \sum_{j=1}^{n} \frac{\epsilon_c}{100} \cdot h_j \qquad (4\text{-}5)$$ where h_j = thickness of each stratum j in in.

b. Dry Sand

Step	Description
1–6	Repeat steps 1 through 6 in Table 4-1a to evaluate D_r and $(N_1)_{60}$.
7	Evaluate mean effective pressure σ_m' of each stratum in tsf (e.g., $\sigma_m' = (1 + 2K_o)/3 \cdot \sigma_o' = 0.65\sigma_o'$ if the coefficient of lateral earth pressure $K_o = 0.47$). σ_m' is considered the total mean pressure in dry sand.
8	Calculate $$G_{max} = 10 \cdot [(N_1)_{60}]^{1/3} \cdot [\sigma_m']^{1/2} \qquad (4\text{-}6)$$ where G_{max} = maximum shear modulus in tsf.
9	Evaluate using G_{max} from step 8 $$\gamma_{eff} \cdot \frac{G_{eff}}{G_{max}} = \frac{0.65 \cdot a_{max}\, \sigma_o \cdot r_d}{g \cdot G_{max}} \qquad (4\text{-}7)$$ where γ_{eff} = effective cyclic shear strain induced by an earthquake; G_{eff} = effective shear modulus at earthquake-induced shear stress in tsf; a_{max} = maximum horizontal acceleration at the ground surface in units of **g**; σ_o = total overburden pressure in tsf; **g** = acceleration of gravity, 32 ft/sec/sec; and r_d = stress reduction factor, 1.0 at the ground surface decreasing to 0.9 at a depth 30 ft below ground surface.
10	Evaluate γ_{eff} from Figure 4-4 using $\gamma_{eff} \cdot G_{eff}/G_{max}$ from step 9 and σ_m' from step 7; multiply by 100 to convert to percent.
11	Use γ_{eff} and evaluate volumetric strain in percent $\epsilon_{c,7.5}$ from Figure 4-5 using D_r or $(N_1)_{60}$ for an M = 7.5-magnitude earthquake.
12	Evaluate volumetric strain ratio from Figure 4-6 for the given magnitude of earthquake M and multiply this ratio by $\epsilon_{c,7.5}$ to calculate $\epsilon_{c,M}$.
13	Multiply $\epsilon_{c,M}$ by 2 to consider the multidirection effect of earthquake shaking on settlement and evaluate total earthquake-induced settlement of each stratum j of thickness h_j for n strata by $$\rho_e = \sum_{j=1}^{n} \frac{2\epsilon_{c,M}}{100} \cdot h_j \qquad (4\text{-}8)$$

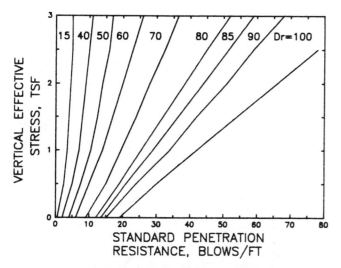

Figure 4-1. Correlations between relative density and blow count N from SPT after Gibbs and Holtz (data from NAVFAC DM-7.1)

face from earthquake records (e.g., Regulation Guide 1.6, Nuclear Regulatory Commission, items 33 and 34; Office, Chief of Engineer policy for Corps of Engineer specifications for ground motions is provided by the Earthquake Engineering and Geosciences Division, Geotechnical Laboratory, USAE Waterways Experiment Station).

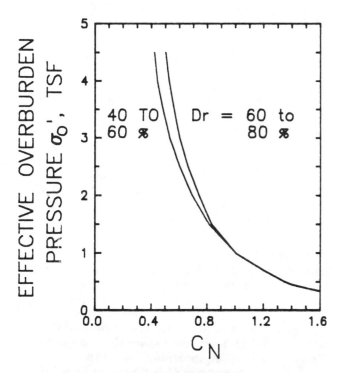

Figure 4-2. Curves for determination of C_N [data from Tokimatsu and Seed 1984)]

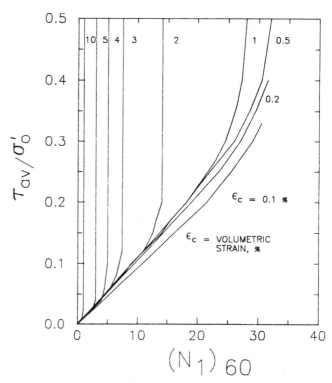

Figure 4-3. Proposed relationship between cyclic stress ratio $(N_1)_{60}$ and volumetric strain ϵ_c for saturated clean sands for M = 7.5 earthquake. Reprinted by permission of the American Society of Civil Engineers from *Journal of Geotechnical Engineering*, Vol 118, 1987, "Evaluation of Settlements in Sands Due to Earthquake Shaking," by K. Tokimatsu and H. B. Seed, p. 866

(1) Application. The procedure is applied to the Tokachioki earthquake in Table 4-2.

(2) Validation. This tentative procedure has not been fully validated. The example problems in Table 4-2 are based on estimated field behaviors and not on measured data against which to validate a settlement analysis.

B. WES PROCEDURE FOR SANDS. The Waterways Experiment Station is currently preparing a procedure on a validated (against centrifuge test data) 2-D soil-structure interactive nonlinear dynamic effective stress analysis that computes dynamic response histories of motions, stresses, pore water pressures, and volume changes for the range of responses and pore water pressures up to and including the initial liquefaction condition. The effects of pore water pressures on moduli, motions, stresses, and volume changes are taken into account for the entire time history of an earthquake.

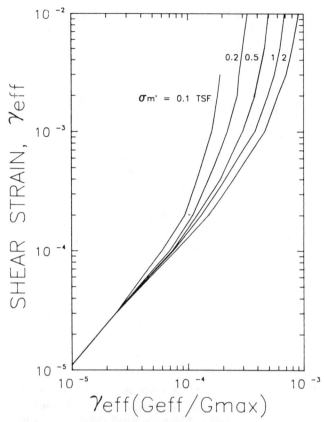

Figure 4-4. Plot for determination of induced strain in sand deposits. Reprinted by permission of the American Society of Civil Engineers from *Journal of Geotechnical Engineering*, Vol 118, 1987, "Evaluation of Settlements in Sands Due to Earthquake Shaking," by K. Tokimatsu and H. B. Seed, p. 873

4-3. Settlement from Repeated Loads and Creep

Structures subject to repeated vertical loads experience a long-term settlement from the compression of cumulative cyclic loads and secondary compression or creep. Operating machinery, pile driving, blasting, and wave or wind action are common causes of this type of dynamic loading. Methods of estimating secondary compression are provided in Section IV, Chapter 3.

A. COMPACTION SETTLEMENT FROM MACHINE VIBRATIONS. A procedure to estimate settlement in sand layers from machine vibrations is described in Table 4-3. The procedure is applied to an example in Table 4-4.

B. SETTLEMENT CALCULATED FROM LABORATORY CYCLIC STRAIN TESTS. Drained cyclic triaxial tests may be performed on pervious soil

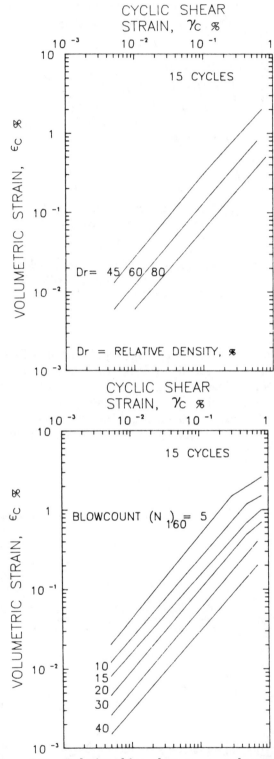

Figure 4-5. Relationships between volumetric strain ϵ_c and cyclic shear strain γ_c for dry sand and earthquake magnitude $M = 7.5$. Reprinted by permission of the American Society of Civil Engineers from *Journal of Geotechnical Engineering*, Vol 118, 1987, "Evaluation of Settlements in Sands Due to Earthquake Shaking," by K. Tokimatsu and H. B. Seed, p. 874

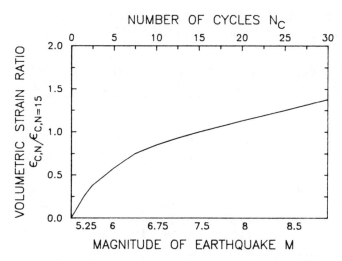

Figure 4-6. Relationship between volumetric strain ratio and number of cycles (earthquake magnitude) for dry sands. Reprinted by permission of the American Society of Civil Engineers from *Journal of Geotechnical Engineering*, Vol 118, 1987, "Evaluation of Settlements in Sands Due to Earthquake Shaking," by K. Tokimatsu and H. B. Seed, p. 874

to evaluate the cyclic settlement through a cyclic strain resistance r_ϵ (item 25).

(1) Test Procedure. The soil should be consolidated to simulate the in situ stress state of effective horizontal and vertical pressures. The soil is subsequently subject to three different cyclic stress levels to evaluate r_ϵ.

(a) The soil should be consolidated so that a plot of one-half of the deviator stress versus the effective horizontal confining pressure provides a slope indicative of a realistic effective coefficient of lateral earth pressure. The slope s of this curve required to obtain a given coefficient of lateral earth pressure K_o is

$$s = 0.5 \cdot \left(\frac{1}{K_o} - 1\right) \qquad (4\text{-}9)$$

For example, the slope s should be 0.7 if K_o is 0.42. The soil should be consolidated to an effective horizontal confining pressure simulating the in situ soil.

(b) Additional vertical dynamic loads should be applied so that the soil specimen is subject to three different cyclic stress levels of 200 to 300 cycles per stress level. The effective lateral confining pressure is kept constant.

(c) The cumulative strain as a function of the number of cycles N at each stress level should be plot-

ted as shown in Figure 4-7a. The slope of the curves in Figure 4-7a is the strain resistance $R_\epsilon = dN/d\epsilon$.

(d) The strain resistance should be plotted versus the number of cycles as shown in Figure 4-7b for each stress level. A straight line should subsequently be plotted through these data points for each stress level. The slope of this line is the cyclic strain resistance r_ϵ.

(e) The cyclic strain resistance decreases with increasing stress levels and approaches zero when the shear strength is fully mobilized. The cyclic strain resistance may increase with increasing depth because the percentage of mobilized shear strength may decrease with increasing depth.

(2) Calculation of Settlement. The settlement of a pervious layer of thickness H caused by repeated loads may be given for this drained soil by (item 25)

$$\rho_r - \frac{H}{r_\epsilon} \cdot \ln N \qquad (4\text{-}17)$$

where ρ_r = settlement of the layer from repeated load in ft; H = thickness of stratum in ft; r_ϵ = cyclic strain resistance of stratum from laboratory tests; and N = number of cycles of repeated load.

The appropriate value of r_ϵ to select from the laboratory test results depends on the maximum anticipated stress level in the soil caused by the repeated loads. For example, the maximum anticipated stress in the soil level may be calculated from the exciting force by methods in Appendix C. The exciting force may be calculated from guidance provided in NAVFAC DM-7.3.

(3) Alternative Settlement Calculation. An alternative method of evaluating effects of repeated loads on settlement of clayey soil from laboratory cyclic triaxial tests is to apply the creep strain rate formulation (item 24)

$$\epsilon_t - \epsilon_{t1} + \frac{e^a}{1 - \lambda_d} \cdot [t^{1-\lambda_d} - t_1^{1-\lambda_d}] \qquad (4\text{-}18a)$$

If $\lambda_d = 1$, then

$$\epsilon_t = \epsilon_{t1} + e^a \cdot \ln\frac{t}{t_1} \qquad (4\text{-}18b)$$

where ϵ = strain at time t; ϵ_{t1} = strain at time t_1 or after one cycle; e = base e or 2.7182818; $\alpha = C \cdot \sigma_{rd} - B$; B, C = constants from Table 4-5; σ_{rd} = repeated deviator stress in tsf; λ_d = decay constant found from slope of logarithmic strain rate ϵ_N/ϵ_1 versus logarithm number of cycles N_c (Figure 4-8).

(a) Settlement may be found by substituting ϵ_t of Equation 4-18b for ϵ_c of Equation 4-5 (Table 4-1a).

Table 4-2a. Example Applications of Simplified Procedure to Estimate Earthquake Settlement: Saturated Sand at Initial Liquefaction Condition (Tokimatsu and Seed 1987)

Layer (1)	Thickess (ft) (2)	N (3)	C_{ER} (4)	σ'_o (psf) (5)	C_N (6)	$(N_1)_{60}$ (7)	$\dfrac{\tau_{av}}{\sigma_o}$ (8)	ϵ_c (percent) (9)	ρ_e (in.) (10)
1	4.0	1.0	0.82[a]	240					
2	3.3	0.5	0.82[a]	575	1.7	0.7	0.155	10.0	4.0
3	3.3	0.5	0.82[a]	764	1.57	0.6	0.185	10.0	4.0
4	3.3	0.5	1.09	954	1.44	0.8	0.200	10.0	4.0
5	3.3	2.0	1.09	1144	1.34	2.9	0.210	5.5	2.2
6	3.3	5.0	1.09	1334	1.24	6.8	0.215	3.2	1.3
7	3.3	23.0	1.21	1523	1.16	32.0	0.220	0.0	0.0
8	3.3	33.0	1.21	1713	1.09	44.0	0.225	0.0	—
9	3.3	28.0	1.21	1903	1.03	35.0	0.225	0.0	—
10	3.3	33.0	1.21	2093	0.97	39.0	0.225	0.0	—

[a]Corrected by 0.75.

Note: Total settlement = 15.5 Water table = 4 ft; Observed maximum settlement $\approx$ 20; Estimated maximum acceleration a_{max} = 0.2 **g**. Reprinted by permission of the American Society of Civil Engineers *Journal of Geotechnical Engineering*, Vol. 118, 1987 "Evaluation of Settlements in Sands Due to Earthquake Shaking," by K. Tokimatsu and H. B. Seed, p. 871.

Table 4-2b. Example Applications of Simplified Procedure to Estimate Earthquake Settlement: Dry Sand (Tokimatsu and Seed 1987)

Layer (1)	Thickness (ft) (2)	σ'_o (psf) (3)	G_{max} (ksf) (4)	$\gamma_{eff}\dfrac{G_{eff}}{G_{max}}$ (5)	γ_{eff} (6)	$\epsilon_{c,7.5}$ percent (7)	$\epsilon_{c,6.6}$ percent (8)	ρ_e (in.) (9)
1	5	240	520	0.00013	0.0005	0.14	0.11	0.13
2	5	715	900	0.00023	0.0008	0.23	0.18	0.22
3	10	1425	1270	0.00032	0.0012	0.35	0.28	0.67
4	10	2375	1630	0.00040	0.0014	0.40	0.32	0.77
5	10	3325	1930	0.00045	0.0015	0.45	0.36	0.86
6	10	4275	2190	0.00046	0.0013	0.38	0.30	0.71

Note: Total estimated settlement = 2.7 in; Total settlement = 3.37 D_r = 45%; $(N_1)_{60}$ = 9; a_{max} = 0.45 **g**. Reprinted by permission of the American Society of Civil Engineers from *Journal of Geotechnical Engineering*, Vol. 118, 1987, "Evaluation of Settlements in Sands Due to Earthquake Shaking," by K. Tokimatsu and H. B. Seed, p. 876.

Table 4-3. Settlement from Machine Vibrations (After NAVFAC DM-7.3)

Step (1)	Description (2)
1	Evaluate initial relative density D_{roj} of each soil layer j from the blowcount N by Equations 4-1, Figure 4-1, or improved correlations (item 50).
2	Estimate or measure maximum displacement of vibration $Amax$ in in. and frequency of vibration f in revolutions per minute at the base of the foundation.
3	Calculate the frequency of vibration in radians per second from $$\omega_o = \frac{2\pi f}{60} \qquad (4\text{-}10)$$
4	Calculate acceleration of vibrations in **g** at foundation level a_o $$a_o = \frac{\omega_o^2 \cdot Amax}{12 \cdot 32.2} \qquad (4\text{-}11)$$
5	Calculate a_j acceleration of vibration in **g** at midpoint of each soil layer j by $$d_{mj} > R: \qquad a_j = a_o \cdot \left[\frac{R}{d_m}\right]^{1/2} \qquad (4\text{-}12a)$$ $$d_{mj} \leq R: \qquad a_j = a_o \qquad (4\text{-}12b)$$ where d_{mj} = distance from foundation base to midpoint of soil layer j in ft; and R = equivalent radius of foundation $\sqrt{LB/\pi}$ in ft.
6	Calculate the critical acceleration in **g** of each soil layer j $$a_{critj} = \frac{-1n \cdot \left(1 - \frac{D_{roj}}{100}\right)}{\beta_v} \qquad (5\text{-}13)$$ where D_{roj} = initial relative density at zero acceleration of layer j in percent; and β_v = coefficient of vibratory compaction. β_v depends on water content W in percent and varies approximately $$W < 5 \text{ percent:} \qquad \beta_v = 0.2 + 0.12W$$ $$5 \leq W \leq 18 \text{ percent:} \qquad \beta_v = 0.77 + 0.006W$$ β_v decreases if the water content is greater than 18 percent.
7	Estimate the final relative density D_{rfj} of each layer j from $$a_j > a_{critj} \qquad D_{rfj} = 100 \cdot [1 - e^{-\beta_v \cdot (a_{critj} + a_j)}] \qquad (4\text{-}14a)$$ $$a_j \leq a_{critj} \qquad D_{rfj} = D_{roj} \qquad (4\text{-}14b)$$
8	Calculate the change in relative density D_{jr} of each soil layer j by $$\Delta D_{rj} = D_{rfj} - D_{roj} \qquad (4\text{-}15)$$
9	Calculate the settlement in ft of each soil layer j by $$\rho_{vj} = 0.0025 \cdot \frac{\Delta D_{rj}}{100} \cdot \gamma_{do} \cdot H_j \qquad (4\text{-}16)$$ where γ_{do} = initial dry density of the sand layer in lbs/ft³; and H_j = stratum of thickness in ft. Equation 4-16 is based on the range of maximum and minimum dry densities for sands reported in item 6.
10	Add the settlements of each layer to find the total settlement.

Table 4-4. Example Calculation for Vibration-Induced Compaction Settlement Under Operating Machinery (From NAVFAC DM-7.3)

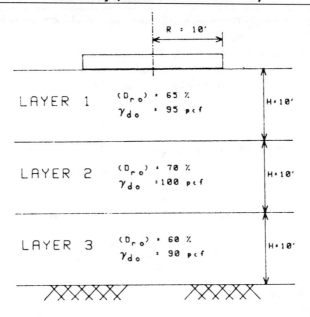

GIVEN: SOIL PROFILE AS SHOWN, FOOTING WITH A RADIUS OF 10 FEET SUBJECTED TO A VIBRATORY LOAD CAUSING A PEAK DYNAMIC DISPLACEMENT $A_{mas} = 0.007$ IN. OPERATING FREQUENCY $f = 2500$ RPM. WATER CONTENT $W = 16$ %, USE $\beta_v = 0.88$

$$\omega_o = \frac{2500}{60} \cdot 2\pi = 261.8 \text{ RAD/SEC}$$

$$a_o = \frac{(261.8)^2 \cdot 0.007}{12 \cdot 32.2} = 1.24g$$

LAYER 1: DEPTH TO MID LAYER $d_m = 5' < R = 10'$

USE $a_j = a_o = 1.24g$

$$a_{critj} = \frac{-\ln\left[1 - \dfrac{D_{roj}}{100}\right]}{\beta_v}$$

$$a_{critj} = \frac{-\ln\left[1 - \dfrac{65}{100}\right]}{0.88} = 1.19g$$

$a_j = 1.24g > a_{critj}$

USE $D_{rfj} = 100\left[1 - e^{-\beta_v \cdot \left[a_{critj} + a_j\right]}\right]$

$\qquad = 100\left[1 - e^{-0.88 \cdot (1.19 + 1.24)}\right]$

$\qquad = 100 \cdot (1 - .118) = 88$ %

$\Delta D_{rj} = D_{rfj} - D_{roj} = 88 - 65 = 23$ %

$$\rho_{vj} = 0.0025 \cdot \frac{\Delta D_{rj}}{100} \cdot \gamma_{do} \cdot H_j$$

$\qquad = 0.0025 \cdot 0.23 \cdot 95 \cdot 10 = 0.55$ FT OR 6.6 IN.

LAYER 2: DEPTH TO MIDLAYER $d_m = 15'$

$$a_j = a_o \cdot \left[\frac{R}{d_m}\right]^{1/2} = 1.24 \cdot \left[\frac{10}{15}\right]^{1/2} = 1.01g$$

$$a_{critj} = \frac{-\ln(1 - 0.7)}{0.88} = 1.37g$$

$a_i = 1.01g < a_{critj}$

$D_{rj} = D_{roj} = 70$ %

NO SIGNIFICANT COMPACTION SETTLEMENT LIKELY

LAYER 3: DEPTH TO MID LAYER $d_m = 25'$

$$a_i = 1.24 \cdot \left[\frac{10}{25}\right]^{1/2} = 0.78g$$

$$a_{critj} = \frac{-\ln(1 - 0.6)}{0.88} = 1.04g$$

$a_j < a_{critj}$

NO SIGNIFICANT COMPACTION SETTLEMENT LIKELY

ANTICIPATED COMPACTION SETTLEMENT = 6.6 IN. RELATIVE DENSITY OF TOP LAYER SHOULD BE INCREASED TO 70 % OR GREATER

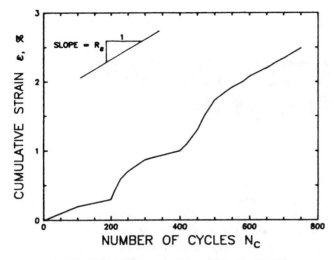

a. CUMULATIVE STRAIN VERSUS NUMBER OF CYCLES

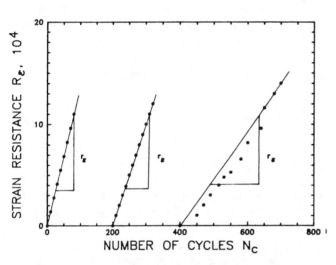

b. STRAIN RESISTANCE VERSUS NUMBER OF CYCLES

Figure 4-7. Example of strain and strain resistance as a function of cycles N_c for different stress levels

Evaluation of ϵ_t from Equations 4-18 is appropriate for repeated loads with frequencies between 0.1 and 10 Hz, a typical range for traffic loads; however, settlement may be underestimated because traffic loads are more complex than compressive vertical loads. Repeated loads with various periods and rest intervals

Table 4-5. Constants B and C to Evaluate Creep Constant α as a Function of Overconsolidation Ratio OCR (Data from Item 24)

OCR (1)	C (2)	B (3)
4	3.5	9.5
10	2.8	9.2
20	3.7	9.5

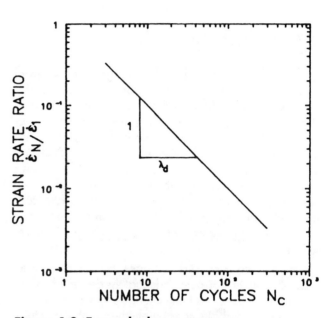

Figure 4-8. Example decay constant

between repeated loads do not appear to cause significant change in strain.

(b) An application of Equations 4-18 to London clay where $\lambda_d = 1$, $\epsilon_{t1} = 0.0$ at $t_1 = 1$ second, $\sigma_{rd} = 1$ tsf, and OCR = 4 is

$$\epsilon_t = \epsilon_{t1} + e^{\alpha} \cdot 1n \frac{t}{t_1} = e^{(3.5 \cdot 1.0 - 9.5)} \cdot 1n\, t$$

$$= \frac{1n\, t}{e^6} = 1n\, t/403.4 = 0.0025\ 1n\, t$$

After 10 seconds the strain ϵ_{10} is 0.0058. Settlement is the strain times thickness of the stratum.

CHAPTER 5

APPLICATIONS WITH UNSTABLE FOUNDATION SOIL

5-1. Unstable Soils

Many types of soils change volume from causes different from elastic deformation, consolidation, and secondary compression. These volume changes cause excessive total and differential movements of overlying structures and embankments in addition to load-induced settlement of the soil. Such unstable conditions include the heaving of expansive clays and collapse of silty sands, sandy silts, and clayey sands from alteration of the natural water content. Refer to Chapter 6 for coping with movements.

A. EFFECTS OF EXCESSIVE MOVEMENTS. Excessive total and, especially, differential movements have caused damages to numerous structures that have not been adequately designed to accommodate the soil volume changes. Types of damage include impaired functional usefulness of the structure, external and interior cracked walls, and jammed and misaligned doors and windows. Important factors that lead to damage are the failure to recognize the presence of unstable soil and to make reasonable estimates of the magnitude of maximum heave or settlement/collapse. Adequate engineering solutions such as special foundation designs and soil-stabilization techniques exist to accommodate the anticipated soil movement. A thorough field investigation is necessary to properly assess the potential movement of the soil. A qualitative estimate of potential vertical movement of proposed new construction may sometimes be made by examination of the performance of existing structures adjacent to the new construction.

B. INFLUENCE OF TIME ON MOVEMENT. The time when heave or settlement/collapse occurs cannot be easily predicted because the location and time when water becomes available to the foundation soil cannot readily be foreseen. Heave or settlement can occur almost immediately in relatively pervious foundation soil, particularly in local areas subject to poor surface drainage and in soil adjacent to leaking water lines. More often, heave or settlement will occur over months or years depending on the availability of moisture. Soil movement may be insignificant for many years following construction, permitting adequate performance until some change occurs in field conditions to disrupt the moisture regime. Predictions of when heave or settlement occurs is usually of little engineering significance. Important engineering problems include reliable determination of the magnitude of potential heave or settlement and development of ways to minimize this potential for movement and potential distress of the structure.

Section I. Heaving Soil

5-2. General

Expansive or swelling soils are found in many areas throughout the United States and the world. These soils change volume within the active zone for heave from changes in soil moisture. Refer to TM 5-818-7, Foundations in Expansive Soil, for details on mechanisms of heave, analysis and design of foundations in expansive soil.

A. SOILS SUSCEPTIBLE TO HEAVE. These soils consist of plastic clays and clay shales that often contain colloidal clay minerals such as the montmorillonites or smectite. They include marls, clayey siltstone and sandstone, and saprolites. Some soils, especially dry residual clayey soil, may heave on wetting under low applied pressure but collapse at higher pressure. Other clayey soil may initially collapse on wetting but heave over long periods of time as water slowly wets the less pervious clay particles. Desiccation can cause expansive soil to shrink.

B. DEPTH OF ACTIVE ZONE. The depth of the active zone Z_a illustrated in Figure 5-1 is defined as the least soil depth above which changes in water content and soil heave may occur because of changes in environmental conditions following construction. The water content distribution should not change with time below Z_a. Experience indicates Z_a may be approximated following guidelines in Table 5-1.

C. EQUILIBRIUM PORE WATER PRESSURE PROFILE. The pore water pressure beneath the center of the foundation is anticipated to reach an equilibrium distribution, whereas the pore water pres-

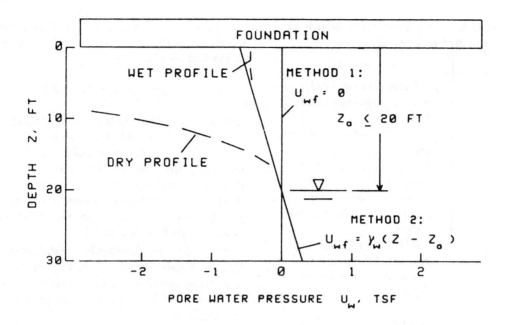

a.   SHALLOW GROUNDWATER LEVEL

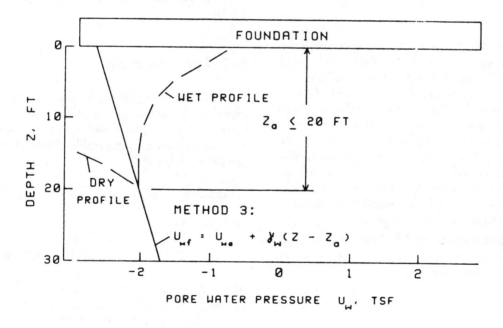

b.   DEEP GROUNDWATER LEVEL

Figure 5-1. Anticipated equilibrium pore water pressure profiles

Table 5-1. Guidelines for Estimating Depth of the Active Zone Z_a

Relative to (1)	Guideline (2)
Water table	Z_a will extend to depths of shallow groundwater levels $\leq$ 20 ft.
Swell pressure	Z_a will be located within depths where $\sigma_{sj} - \sigma_{fj} \geq 0$, $\sigma_{sj} =$ average swell pressure of stratum j, and $\sigma_{fj} =$ total average vertical overburden pressure after construction in stratum j.
Fissures	Z_a will be within the depth of the natural fissure system caused by seasonal swell/shrinkage.
Climate	
Humid	$Z_a = 10$ ft.
Semiarid	$Z_a = 15$ ft.
Arid	$Z_a = 20$ ft.

sure profile beneath the perimeter will cycle between dry and wet extremes, depending on the availability of water and the climate. Placement of a foundation on the soil may eliminate or reduce evaporation of moisture from the ground surface and eliminate transpiration of moisture from previously existing vegetation. Figure 5-1 illustrates three methods described later for estimating the equilibrium pore water pressure profile u_{wf} in tsf. If undisturbed soil specimens are taken from the field near the end of the dry season, then the maximum potential heave may be estimated from results of swell tests performed on these specimens.

(1) Saturated Profile (Method 1, Figure 5-1). The equilibrium pore water pressure in the saturated profile within depth Z_a is

$$u_{wf} = 0 \qquad (5\text{-}1a)$$

This profile is considered realistic for most practical cases, including houses or buildings exposed to watering of perimeter vegetation and possible leaking of underground water and sewer lines. Water may also condense or collect in permeable soil beneath foundation slabs and penetrate into underlying expansive soil unless drained away or protected by a moisture barrier. This profile should be used if other information on the equilibrium pore water pressure profile is not available.

(2) Hydrostatic with Shallow Water Table (Method 2, Figure 5-1). The equilibrium pore water pressure in this profile is zero at the

groundwater level and decreases linearly with increasing distance above the groundwater level in proportion to the unit weight of water

$$u_{wf} = \gamma_w(z - Z_a) \qquad (5\text{-}1b)$$

where $\gamma_w =$ unit weight of water, 0.031 ton/ft^3; and $z =$ depth below the foundation in ft.

This profile is considered realistic beneath highways and pavements where surface water is drained from the pavement and underground sources of water such as leaking pipes or drains do not exist. This assumption leads to smaller estimates of anticipated heave than Method 1.

(3) Hydrostatic Without Shallow Water Table (Method 3, Figure 5-1). The pore water pressure of this profile is similar to Method 2 but includes a value of the negative pore water pressure u_{wa} at depth Z_a.

$$u_{wf} = u_{wa} + \gamma_w(z - Z_a) \qquad (5\text{-}1c)$$

u_{wa} may be evaluated by methods described in TM 5-818-7.

5-3. Identification

Soils susceptible to swelling can be most easily identified by simple classification tests such as Atterberg limits and natural water content. Two equations that have provided reasonable estimates of free swell are (items 30, 64)

$$\log S_f = 0.0367LL + 0.0833W_n + 0.458 \qquad (5\text{-}2a)$$

and (item 49)

$$S_f = 2.27 + 0.131LL - 0.27W_n \qquad (5\text{-}2b)$$

where $S_f =$ free swell in percent; $LL =$ liquid limit in percent; and $W_n =$ natural water content in percent.

The percent swell under confinement can be estimated from the free swell by (item 20)

$$S = S_f \cdot (1 - 0.72\sqrt{\sigma_f}) \qquad (5\text{-}3)$$

where $S =$ swell under confinement in percent; and $\sigma_f =$ vertical confining pressure in tsf.

These identification procedures were developed by correlations of classification test results with results of 1-D swell tests performed in consolidometers on undisturbed and compacted soil specimens. Soils with liquid limit less than 35 percent and plasticity index less than 12 percent have relatively low potential for swell and may not require swell testing. Refer to TM

5-818-7 for further details on identification of expansive soils.

5-4. Potential Vertical Heave

Useful estimates of the anticipated heave based on results from consolidometer swell tests can often be made.

A. SELECTION OF SUITABLE TEST METHOD. Suitable standard test methods for evaluating the potential for 1-D heave or settlement of cohesive soils are fully described in EM 1110-2-1906 and ASTM D 4546. A brief review of three 1-D consolidometer tests useful for measuring potential swell or settlement using a standard consolidometer illustrated in Figure E-1, Appendix E, is provided herein.

(1) Free Swell. After a seating pressure (e.g., 0.01 tsf applied by the weight of the top porous stone and load plate) is applied to the specimen in a consolidometer, the specimen is inundated with water and allowed to swell vertically until primary swell is complete. The specimen is loaded following primary swell until its initial void ratio/height is obtained. The total pressure required to reduce the specimen height to the original height prior to inundation is defined as the swell pressure σ_s.

(2) Swell Overburden. After a vertical pressure exceeding the seating pressure is applied to the specimen in a consolidometer, the specimen is inundated with water. The specimen may swell, swell then contract, contract, or contract then swell. The vertical pressure is often equivalent to the in situ overburden pressure and may include structural loads depending on the purpose of the test.

(3) Constant Volume. After a seating pressure and additional vertical pressure, often equivalent to the in situ overburden pressure, is applied to the specimen in a consolidometer, the specimen is inundated with water. Additional vertical pressure is applied as needed or removed to maintain a constant height of the specimen. A consolidation test is subsequently performed as described in Appendix E. The total pressure required to maintain a constant height of the specimen is the measured swell pressure. This measured swell pressure is corrected to compensate for sample disturbance by using the results of the subsequent consolidation test. A suitable correction procedure is similar to that for estimating the maximum past pressure.

B. CALCULATION FROM VOID RATIO. The anticipated heave is

$$S_{max} = \sum_{j=1}^{n} S_{maxj} = \sum_{j=1}^{n} \frac{e_{fj} - e_{oj}}{1 + e_{oj}} \cdot H_i \quad (5\text{-}4a)$$

where S_{max} = maximum potential vertical heave in ft; n = number of strata within the depth of heaving soil; S_{maxj} = heave of soil in stratum j in ft; H_i = thickness of stratum j in ft; e_{fj} = final void ratio of stratum j; and e_{oj} = initial void ratio of stratum j.

The initial void ratio, which depends on a number of factors such as the maximum past pressure, type of soil, and environmental conditions, may be measured by standard consolidometer test procedures described in EM 1110-2-1906 or ASTM D 4546. The final void ratio depends on changes in soil confinement pressure and water content following construction of the structure; it may be anticipated from reasonable estimates of the equilibrium pore water pressure u_{wf}, depth of active zone Z_a, and edge effects by rewriting Equation 5-4a in terms of swell pressure shown in Equation 5-4b.

C. CALCULATION FROM SWELL PRESSURE. The anticipated heave in terms of swell pressure is

$$S_{max} = \sum_{j=1}^{n} \frac{C_{sj}}{1 + e_{oj}} \cdot \log_{10} \frac{\sigma_{sj}}{\sigma'_{fj}} \cdot H_i \quad (5\text{-}4b)$$

where C_{sj} = swell index of stratum j; σ_{sj} = swell pressure of stratum j in tsf; σ'_{fj} = final or equilibrium average effective vertical pressure of stratum j, $\sigma_{fj} - u_{wfj}$ in tsf; σ_{fj} = final average total vertical pressure of stratum j in tsf; and u_{wfj} = average equilibrium pore water pressure in stratum j in tsf.

The number of strata n required in the calculation is that observed within the depth of the active zone for heave.

(1) Swell Index. The swell or rebound index of soil in each stratum may be determined from results of consolidometer tests as described in Section III, Chapter 3, and Figure 3-13. Preliminary estimates of the swell index may be made from Figure 3-14.

(a) The swell index C_s measured from a swell overburden test (swell test described in EM 1110-2-1906 or method B described in ASTM D 4546) may be less than that measured from a constant-volume test (swell pressure test described in EM 1110-2-1906 or method C described in ASTM D 4546). The larger values of C_s are often more appropriate for analysis of potential heave and design.

(b) A simplified first approximation of C_s developed from Corps of Engineer project sites through central Texas is $C_s \approx 0.03 + 0.002(LL-30)$.

(2) Swell Pressure. The swell pressure of soil in each stratum may be found from results of consolidometer swell tests on undisturbed specimens as described in EM 1110-2-1906 or ASTM D 4546. Preliminary estimates of swell pressure may be made from

(item 32)

$$\log_{10}\sigma_s = \overline{2}.1423 + 0.0208LL$$

$$+ 0.0106\gamma_d - 0.0269W_n \quad (5\text{-}5a)$$

where σ_s = swell pressure in tsf; and γ_d = dry density in lbs/ft³.

An alternative equation (item 46) is

$$\sigma_s = 0.00258 \cdot PI^{1.12} \cdot \left[\frac{C}{W_n}\right]^2 + 0.273 \quad (5\text{-}5b)$$

where PI = plasticity index in percent; and C = clay content in percent less than 2 microns.

(3) Final Effective Vertical Pressure.
The final total pressure σ_f may be estimated from the sum of the increase in soil stresses from the structural loads calculated by methods in Appendix C or Figure 1-2 and the initial overburden pressure σ_o. The final effective pressure σ_f' is σ_f less the assumed equilibrium pore water pressure profile u_{wf} (Figure 5-1).

5-5. Potential Differential Heave

Differential heave results from edge effects beneath a finite covered area, drainage patterns, lateral variations in thickness of the expansive foundation soil, and effects of occupancy. The shape, geometry, and loads of the structure also promote differential movement. Examples of the effect of occupancy include broken or leaking underground water lines and irrigation of vegetation adjacent to the structure. Other causes of differential heave include differences in distribution of loads and footing sizes.

A. PREDICTABILITY OF VARIABLES.
Reliable estimates of the anticipated differential heave and location of differential heave are not possible because of uncertainty in such factors as future availability of moisture, horizontal variations in soil parameters, areas of soil wetting, and effects of future occupancy.

B. MAGNITUDE OF DIFFERENTIAL HEAVE. The difference in potential heave between locations beneath a foundation can vary from zero to the maximum potential vertical heave. Differential heave is often the anticipated total heave for structures on isolated spot footings or drilled shafts because soil beneath some footings or portions of slab foundations may experience no wetting and no movement. Refer to Chapter 2 for details on effect of differential movement on performance of the foundation.

(1) A reasonable estimate of the maximum differential movement or differential potential heave ΔS_{max} is the sum of the maximum calculated settlement ρ_{max}

of soil beneath a nonwetted point of the foundation and the maximum potential heave S_{max} following wetting of soil beneath some adjacent point of the foundation separated by the distance ℓ. If all of the soil heaves, then ΔS_{max} is the difference between S_{max} and S_{min} between adjacent points where S_{min} is the minimum heave.

(2) The location of S_{max} may be beneath the most lightly loaded portion of the foundation such as beneath the center of the slab.

(3) The location of ρ_{max} may be beneath columns and consist only of immediate elastic settlement ρ_i in soil where wetting does not occur or will be S_{min} if wetting does occur in expansive soil.

(4) The deflection ratio is $\Delta S_{max}/L$ where L may be the distance between stiffening beams.

5-6. Application

A stiffened ribbed mat is to be constructed on an expansive soil. The soil parameters illustrated in Table 5-2 were determined on specimens of an undisturbed soil sample taken 10 ft beneath the mat. Additional tests at other depths will improve reliability of these calculations. The active zone for heave is estimated to extend 20 ft below ground surface or 20 ft below the base of the mat and 17 ft below the base of the columns. The maximum anticipated heave S_{max} and differential heave S_{max} are to be estimated beneath portions of the mat. Stiffening beams are 3 ft deep with 20-ft spacing in both directions (Figure 5-2). Column loads of 25 tons interior and 12.5 tons perimeter lead to an applied pressure on the column footings $q = 1.0$ tsf. Minimum pressure q_{min} beneath the 5-inch-thick flat slab is approximately 0.05 tsf. The heave calculations assume a zero stiffness mat. Computer program VDISPL in Appendix F is useful for calculating potential heave beneath footings and mat foundations in multi-layered expansive soil. VDISPL also considers heave in an excavation from changes in pore water pressure.

Table 5-2. Soil Parameters for Example Estimation of Anticipated Heave

Parameter (1)	Value (2)
Elastic modulus E_s (tsf)	200
Swell pressure σ_s (tsf)	1.0
Compression index C_c	0.25
Swell index C_s	0.10
Initial void ratio e_o	0.800
Unit wet soil weight γ (ton/ft³)	0.06
Active zone for heave Z_a (ft)	20

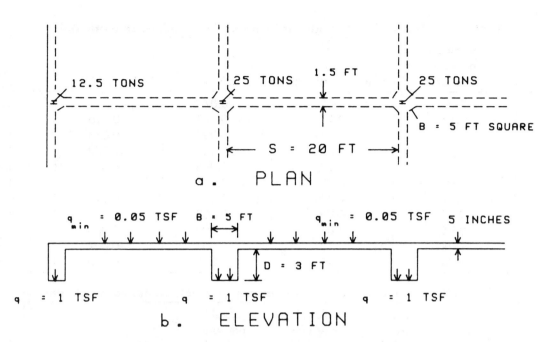

Figure 5-2. Plan and elevation of stiffened mat in expansive soil

A. CALCULATION OF POTENTIAL HEAVE

(1) Maximum Potential Heave S_{max}. The maximum heave is anticipated beneath unloaded portions of the mat. The potential heave is estimated assuming the equilibrium pore water pressure $u_{wf} = 0$ or the soil is saturated; therefore, the final effective pressure $\sigma_f' = \sigma_f$ or the final total pressure.

(a) Table 5-3a illustrates the estimation of anticipated heave S_{max} beneath lightly loaded portions of the mat using Equation 5-4b, increment thickness $\Delta H = 2$ ft, and results of a single consolidometer swell test.

(b) Table 5-3a and Figure 5-3a show that $S_{max} = 0.3$ ft or 3.6 in. and that heave is not expected below 16 ft of depth where the swell pressure approximately equals the total vertical pressure σ_f.

(c) Most heave occurs at depths less than 5 ft below the flat portion of the mat. Replacing the top 4 ft of expansive soil with nonexpansive backfill will reduce S_{max} to 0.115 ft or 1.4 in. (Table 5-3a and Figure 5-3a).

(2) Minimum Potential Heave S_{min}. The minimum potential heave on wetting of the soil to a saturated profile (method 1, Figure 5-1) is expected beneath the most heavily loaded portions of the mat or beneath the columns. Table 5-3b and Figure 5-3b show that the minimum heave S_{min} calculated after Equation 5-4b substituting S_{min} for S_{max} is 0.092 ft beneath the column or S_{min} is 0.092 ft or 1.1 in. beneath the column. Heave is not expected below 13 ft beneath the columns.

B. MAXIMUM DIFFERENTIAL HEAVE ΔS_{max}

(1) ΔS_{max} is the sum of S_{max} and the immediate settlement ρ_i if soil wetting is nonuniform. The maximum immediate settlement ρ_i is anticipated to occur as elastic settlement beneath the loaded columns if soil wetting does not occur in this area. A common cause of nonuniform wetting is leaking underground water lines. From the improved Janbu approximation (Equation 3-17 and Figure 3-8) with reference to Figure 5-2

$$\mu_o = 0.92 \text{ for } D/B = 1.0$$

$$\mu_1 = 0.7 \text{ for } L/B = 1.0 \text{ and } H/B > 10$$

$$\rho_i = \mu_o \cdot \mu_1 \cdot \left[\frac{q \cdot B}{E_s}\right] = 0.92 \cdot 0.7 \cdot \frac{1.0 \cdot 5.0}{200}$$

The maximum differential heave $\Delta S_{max} = S_{max} + \rho_i = 3.6 + 0.2 = 3.8$ in. or 0.317 ft. The deflection ratio Δ/L is $\Delta S_{max}/L = 0.317/20$ or 1/64 where L is 20 ft, the stiffening beam spacing. This deflection ratio cannot be tolerated (Chapter 2). If the top 6 ft of expansive soil is replaced with nonexpansive backfill $\Delta S_{max} = 0.063 + 0.016 = 0.079$ ft or 0.95 in. Ribbed mat foundations and superstructures may be designed to accommodate differential heave of 1 in. after methods in TM 5-818-7 or item 28.

(2) ΔS_{max} is the difference between S_{max} and S_{min} if soil wetting occurs beneath the columns or 3.6 −

Table 5-3a. Heave Calculations for Example Application: Beneath Slab

Depth z (ft) (1)	Overburden pressure $\sigma_o = \gamma z$ (tsf) (2)	Total pressure $\sigma_f = \sigma'_f$ (tsf) (3)	$\frac{S_{maxj}}{\Delta H}$ (4)	S_{maxj} (ft) (5)	S_{max} (ft) (6)
0	0.00	0.05	0.072		0.303
2	0.12	0.17	0.043	0.115	0.188
4	0.24	0.29	0.030	0.073	0.115
6	0.36	0.41	0.022	0.052	0.063
8	0.48	0.53	0.015	0.037	0.026
10	0.60	0.65	0.010	0.025	0.001
12	0.72	0.77	0.006	0.016	−0.015
14	0.84	0.89	0.003	0.009	−0.024
16	0.96	1.01	0.000	0.003	−0.027
18	1.08	1.13	−0.007	−0.007	−0.020
20	1.20	1.25	−0.013	−0.020	0.000

Table 5-3b. Heave Calculations for Example Calculation: Beneath Columns

Depth z (ft)	$\frac{z}{B}$ (2)	Overburden pressure $\sigma_o = \gamma z$ (tsf) (3)	Column* pressure $\Delta \sigma_z$ (tsf) (4)	Total pressure $\sigma_f = \sigma'_f$ (tsf) (5)	$\frac{S_{minj}}{\Delta H}$ (6)	S_{minj} (ft) (7)	S_{min} (ft) (8)
0	0.0	0.00	1.00	1.00	0.000		0.092
1	0.2	0.06	0.96	1.02	−0.006	−0.003	0.095
3	0.6	0.18	0.61	0.79	0.006	0.000	0.095
5	1.0	0.30	0.30	0.60	0.012	0.018	0.077
7	1.4	0.42	0.20	0.62	0.012	0.024	0.053
9	1.8	0.54	0.16	0.70	0.009	0.021	0.032
11	2.2	0.66	0.09	0.75	0.007	0.016	0.016
13	2.6	0.78	0.07	0.85	0.004	0.011	0.005
15	3.0	0.90	0.05	0.95	0.001	0.005	0.000
17	3.4	1.02	0.04	1.06	−0.001	0.000	0.000

* Increase in pressure beneath columns calculated from Figure 1-2, Table C-1a (point under corner rectangular area) where $\Delta \sigma_z = 4q$ or Figure C-2 where $\Delta \sigma_z = 4q$.

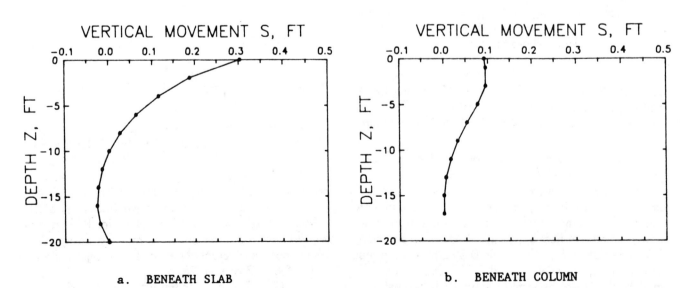

a. BENEATH SLAB   b. BENEATH COLUMN

Figure 5-3. Calculated heave profile beneath mat foundation

1.4 = 2.2 in. Replacement of the top 4 ft of soil beneath the ribbed mat will reduce this differential heave to about 1.4 − 1.1, or about 0.3 in. ignoring the difference in settlement beneath the fill and original expansive soil within 1 ft beneath the column.

Section II. Collapsible Soil

5-7. General

Many collapsible soils are mudflows or wind-blown silt deposits of loess often found in arid or semi-arid climates such as deserts, but dry climates are not necessary for collapsible soil. Loess deposits cover parts of the western, midwestern, and southern United States, Europe, South America, Asia including large areas of Russia and China, and Southern Africa. A collapsible soil at natural water content may support a given foundation load with negligible settlement, but when water is added to this soil the volume can decrease significantly and cause substantial settlement of the foundation, even at relatively low applied stress or the overburden pressure. The amount of settlement depends on the initial void ratio, stress history of the soil, thickness of the collapsible soil layer, and magnitude of the applied foundation pressure. Collapsible soils exposed to perimeter watering of vegetation around structures or leaking utility lines are most likely to settle. Collapse may be initiated beneath the ground surface and propagate toward the surface leading to sudden and nonuniform settlement of overlying facilities.

A. STRUCTURE. Soils subject to collapse have a honeycombed structure of bulky shaped particles or grains held in place by a bonding material or force illustrated in Figure 5-4. Common bonding agents include soluble compounds such as calcareous or ferrous cementation that can be weakened or partly dissolved by water, especially acidic water. Removal of the supporting material or force occurs when water is added, enabling the soil grains to slide or shear and move into voids.

B. COLLAPSE TRIGGER. Table 5-4 illustrates four types of wetting that can trigger the collapse of soil. Dynamic loading may also cause a shear failure in the bonding material and induce collapse. This mechanism is particularly important for roads, airfields, railways, foundations supporting vibrating machinery, and other foundations subject to dynamic forces.

5-8. Identification

Typical collapsible soils are lightly colored; low in plasticity with liquid limits below 45, plasticity in-

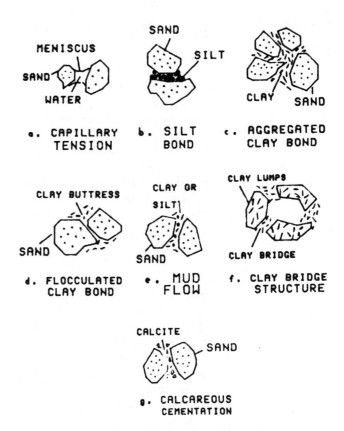

Figure 5-4. Mechanisms for collapse of loose, bulky grains

dices below 25, and relatively low dry densities between 65 and 105 lbs/ft³ (60 to 40 percent porosity). Collapse rarely occurs in soil with a porosity less than 40 percent. Most past criteria for determining the susceptibility of collapse are based on relationships between the void ratio, water content, and dry density (Table 5-5). The methods in Table 5-5 apply to fine-grained soil.

(1) The Gibbs and Bara method (item 18) assumes collapse of soil with sufficient void space to hold the liquid limit water.

(2) Fine-grained soils that are not susceptible to collapse by the criteria in Table 5-5 may have potential for expansion described in Section I of this chapter.

5-9. Potential Collapse

When water becomes available to collapsible soil, settlement and elastic settlement will occur without any additional applied pressure. This settlement will occur quickly in a free draining or pervious soil, but more slowly in a poor draining or less pervious soil. When construction occurs on soil where surface water filters through the collapsible soil over time, some collapse

Table 5-4. Wetting that can Trigger Soil Collapse

Type of wetting (1)	Description (2)
Local, shallow	Wetting of a random nature caused by water sources from pipelines or uncontrolled drainage of surface water; no rise in groundwater level; settlement occurs in upper soil layer within wetted area.
Intense local	Intense deep local wetting caused by discharge of deep industrial effluent, leaking underground utility lines, or irrigation. Flow rates sufficient to cause a continuous rise in groundwater level may saturate the entire zone of collapsible soil within a short time (i.e., < 1 year) and cause uneven and damaging settlement under existing structural loads or only the soil weight.
Slow, uniform rise in groundwater	Slow relatively uniform rise of groundwater from sources outside of the collapsible soil area will cause uniform and gradual settlement.
Slow increase in water content	Gradual increase in water content of thick collapsible soil layer from steam condensation or reduction in evaporation from the ground surface following placement of concrete or asphalt will cause incomplete settlement.

Table 5-5. Relationships for Estimating Susceptibility of Soil to Collapse

Source (1)	Soil susceptible to collapse (2)
Northey 1969 (item 48)	Denisov introduced a coefficient of subsidence $k_d = e_{LL}/e_o$; the soil is collapsible if $$0.5 < k_d < 0.75$$ where e_{LL} = void ratio at liquid limit; LL = liquid limit in percent; and e_o = natural void ratio.
After Gibbs and Bara 1962 (item 18)	$$\gamma_d < 162.3/(1 + 0.026LL)$$ where γ_d = natural dry density in lbs/ft^3, or $e_o > 2.6LL/100$
Feda 1966 (item 16)	$$\frac{\left(\frac{e_o}{100G_s}\right) - PL}{PI} > 0.85$$ where PL = plastic limit in percent; PI = plasticity index in percent; and G_s = specific gravity of soil.
Jennings and Knight 1975 (item 26)	Measure of collapse potential CP of a specimen tested in a one-dimensional consolidometer in percent $$CP = \frac{e_o - e_c}{1 + e_o} \cdot 100$$ where e_o = void ratio at σ = 2 tsf at natural water content prior to wetting; and e_c = void ratio after soaking at $\sigma - 2$ tsf.

CP percent	Severity of Collapse
0–1	negligible
1–5	moderate trouble
5–10	trouble
10–20	severe trouble
>20	very severe trouble

will occur in situ and reduce the collapse that will occur on wetting following construction. Procedures for estimating the potential for collapse are uncertain because no single criterion can be applied to all collapsible soil. The amount of settlement depends on the extent of the wetting front and availability of water, which rarely can be predicted prior to collapse. Laboratory classification and consolidation tests can fail to indicate soil that eventually does collapse in the field. The following procedures to estimate collapse attempt to follow the stress path to which the soil will be subjected in the field. Immediate settlement prior to collapse may be estimated by methods in Sections I and II, Chapter 3.

A. WETTING AT CONSTANT LOAD.
An acceptable test procedure is described in detail as method B of ASTM D 4546 or the swell test procedure in appendix VIIIA of EM 1110-2-1906. A specimen is loaded at natural water content in a consolidometer to the anticipated stress that will be imposed by the structure in the field. Distilled water (or natural site water if available) is added to the consolidometer and the decrease in specimen height following collapse is noted.

The settlement of collapsible soil may be estimated by

$$\rho_{col} = \frac{e_o - e_c}{1 + e_o} \cdot H \qquad (5\text{-}6)$$

where ρ_{col} = settlement of collapsible soil stratum in ft; e_o = void ratio at natural water content under anticipated vertical applied pressure σ_f; e_c = void ratio following wetting under σ_f; and H = thickness of collapsing soil stratum in ft.

The total settlement of the soil will be the sum of the settlement of each stratum.

B. MODIFIED OEDOMETER TEST (ITEM 22). This test is a modification of the Jennings and Knight (item 26) double oedometer procedure that eliminates testing of two similar specimens, one at natural water content and the other inundated with distilled (or natural) water for 24 hrs.

(1) **Procedure.** An undisturbed specimen is prepared and placed in a 1-D consolidometer at the natural water content. The initial specimen height h is recorded. A seating pressure of 0.05 tsf is placed on the specimen and the dial gauge is zeroed (compression at stress levels less than 0.05 tsf is ignored). Within 5 minutes, the vertical stress is increased in increments of 0.05, 0.1, 0.2, 0.4 tsf, etc., until the vertical stress is equal to or slightly greater than that expected in the field following construction. For each increment, dial readings are taken every 1/2 hr until less than 0.1 percent compression occurs in 1 hr. The specimen is subsequently inundated with distilled (or natural) water and the collapse observed on the dial gauge is recorded. Dial readings are monitored every 1/2 hr at this stress level until less than 0.1 percent compression occurs in 1 hr. Additional stress is placed on the specimen in increments as previously described until the slope of the curve is established. The dial readings d are divided by the initial specimen thickness h_o and multiplied by 100 to obtain percent strain. The percent strain may be plotted as a function of the applied pressure as shown in Figure 5-5 and a dotted line projected from point C to point A to approximate the collapse strain for stress levels less than those tested.

(2) **Calculation of Collapse.** The soil profile should be divided into different layers with each layer corresponding to a representative specimen such as the one illustrated in Figure 5-6. The initial and final stress distribution should be calculated for each layer and entered in the compression curve such as in Figure 5-5 and the vertical strain recorded at the natural water content and the inundated water content. The settlement is the difference in strain between the natural water content and wetted specimen at the same stress level

$$\rho_{col} = [(d/h_o)_f - (d/h_o)_o] \cdot H/100 \qquad (5\text{-}7)$$

where ρ_{col} = collapse settlement in ft; $(d/h_o)_f \cdot 100$ = strain after wetting at the field stress level in percent; $(d/h_o)_o \cdot 100$ = strain at natural water content at the field stress level in percent; d = dial reading in in.; h_o = initial specimen height in in.; and H = thickness of collapsible stratum in ft. Total settlement is the sum of the collapse settlement of each stratum.

5-10. Application

A 3-ft-square footing illustrated in Figure 5-6 is to be placed 3-ft deep on a loess soil with a thickness of 5 ft beneath the footing. The results of a modified oedometer test performed on specimens of this soil are provided in Figure 5-5. The footing pressure q = 1 tsf. Refer to Appendix F for calculation of potential collapse of a footing using program VDISPL.

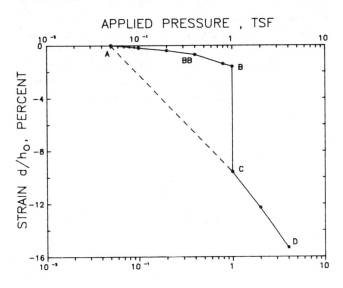

d = DIAL READING, IN.
h_o = INITIAL SPECIMEN HEIGHT, IN.

Figure 5-5. Example Compression curve of the modified oedometer test. (d/h_o is multiplied by 100 to obtain percent)

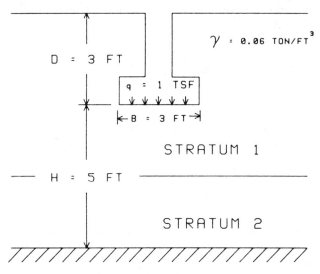

Figure 5-6. Footing for calculating settlement on collapsible soil

Table 5-6a. Example Calculation of Settlement of a Collapsible Soil beneath a Square Footing (Figure 5-6): Stress Distribution

Depth below Footing z (ft) (1)	$\frac{B}{2z}$ (2)	Overburden pressure σ_o (tsf) (3)	Influence Factor[a] I_σ		Footing Stress[b] q_z (tsf)		Total Stress σ_{fz} (tsf)	
			Center (4)	Corner (5)	Center (6)	Corner (7)	Center (8)	Corner (9)
0.0	∞	0.18	0.250	0.250	1.000	0.250	1.180	0.430
2.5	0.6	0.33	0.106	0.195	0.424	0.195	0.754	0.525
5.0	0.3	0.48	0.038	0.106	0.152	0.106	0.632	0.586

[a]From Figure C-2 where $m = n = B/2z$ for the center and $m = n = B/z$ for the corner.
[b]Center: $q_z = 4q \cdot I$; corner: $q_z = q \cdot I$; $q = 1$ tsf.

Table 5-6b. Example Calculation of Settlement of a Collapsible Soil beneath a Square Footing (Figure 5-6): Settlement

Depth below footing z (ft)	Layer (2)	Average Final Stress σ_{fz} (tsf)		Average Initial Strain (Percent)		Average Final Strain (Percent)	
		Center (3)	Corner (4)	Center (5)	Corner (6)	Center (7)	Corner (8)
1.25	1	0.967	0.478	1.55	0.85	9.45	7.30
3.75	2	0.693	0.555	1.25	1.05	8.35	7.75

Settlement from Equation 5-7: Center: $\rho_{col} = [(9.45 - 1.55) + (8.35 - 1.25)] \cdot 2.5 = 0.375$ ft or 4.5 in. Corner: $\rho_{col} = [(7.30 - 0.85) + (7.75 - 1.05)] \cdot 2.5 = 0.329$ ft or 4.0 in.

A. CALCULATION. Tables 5-6a and 5-6b illustrate computation of the vertical stress distribution and collapse settlement at the center and corner of this footing. The stress levels and vertical strains of the soil in Figure 5-5 are shown in Table 5-6b assuming layers 1 and 2 (Figure 5-6) consist of the same soil. The average settlement of (4.5 + 4.0)/2 = 4.3 in. should provide a reasonable estimate of the settlement of this footing.

B. TESTING ERRORS. The amount of collapse depends substantially on the extent of the wetting front and initial negative pore water or suction pressure in the soil, which may not be duplicated because soil disturbance and lateral pressures may not be simulated. Collapse may also be stress path−dependent and may involve a mechanism other than addition of water such as exposure to dynamic forces.

CHAPTER 6

COPING WITH SOIL MOVEMENTS

Section I. Minimizing and Tolerating Soil Movements

6-1. General

Development of society leads increasingly to construction on marginal (soft, expansive, collapsible) soil subject to potential volume changes. Sufficient soil exploration and tests are necessary to provide reliable soil parameters for evaluating reasonable estimates of total and differential settlement.

A. EXPLORATORY BORINGS. Exploratory borings should be made within soil areas supporting the structure and sufficient tests performed to determine up-per and lower limits of the soil strength, stiffness, and other required parameters. Depth of borings should be sufficient to include the significantly stressed zones of soil from overlying structures. These depths should be twice the minimum width of footings or mats with length-to-width ratios less than two, four times the minimum width of infinitely long footings or embankments, or to the depth of incompressible strata, whichever is less.

B. MITIGATION FOR EXCESSIVE DEFORMATION POTENTIAL. If analysis by methods in this manual indicates excessive settlement or heave of the supporting soil, then the soil should be improved and/or various design measures should be applied to reduce the potential volume changes and foundation movements to within tolerable limits.

C. ADDITIONAL REFERENCE. Refer to chapter 16, TM 5-818-1, for further information on stabilization of foundation soil.

6-2. Soil Improvement

Most foundation problems occur from high void ratios, low-strength materials, and unfavorable water content in the soil; therefore, basic concepts of soil improvement include densification, cementation, reinforcement, soil modification or replacement, drainage, and other water content controls. A summary with description of soil improvement methods is shown in Table 6-1. The range of soil particle sizes applicable for these soil improvement methods is shown in Table 6-2. Methods that densify soil by dynamic forces such as vibro-compaction and dynamic compaction (consolidation) may lead to a temporary, short-term reduction in the strength of the foundation soil.

A. SOFT SOIL. Soft soils have poor volume stability and low strength and may be composed of loose sands and silts, wet clays, organic soils, or combinations of these materials. Most of the methods listed in Tables 6-1 and 6-2 are used to minimize settlement in soft soil. Applicability of these methods depends on economy; effectiveness of treatment in the existing soil; availability of equipment, materials, and skills; and the effect on the environment such as disposal of waste materials. Some of the more useful methods for improving soft soil are described in more detail later.

(1) Removal by Excavation. Soft soil underlain by suitable bearing soil at shallow depths (less than 20 ft) may be economical to remove by excavation and replace with suitable borrow material or with the original soil after drying or other treatment. Compacted lean clays and sands (if necessary, with chemical admixtures such as lime, fly ash, or Portland cement) are adequate replacement materials if the water table is below the excavation line. Granular material such as sand, slag, and gravel should be used if the water table is above the bottom of the excavation. Additional mechanical compaction may be accomplished with vibratory or dynamic methods (Table 6-1).

(2) Precompression. Precompression densifies the foundation soil by placing a load or surcharge fill, usually a weight that exceeds the permanent structure load, on the site. The preload should eliminate most of the post-construction primary consolidation and some secondary compression and increase the soil strength.

(a) For embankments, additional fill beyond that required to construct the embankment is usually placed.

(b) For foundations other than earth structures, the preload must be removed prior to construction.

(c) Time required for preload may sometimes be appreciably reduced by sand or prefabricated vertical (PV) strip drains to accelerate consolidation of thick layers of low permeability. PV drains commonly

Table 6-1. Soil Improvement Methods (Includes Data from Item 10)

Method (1)	Principle (2)	Most suitable soils and types (3)	Maximum effective treatment, depth (ft) (4)	Advantages and limitations (5)
a. Vibrocompaction				
Blasting	Shock waves cause liquefaction, displacement, remolding	Saturated clean sands, partly saturated sands, and silts after flooding	60	Rapid, low-cost, treat small areas, no improvement near surface, dangerous
Terraprobe	Densify by vertical vibration, liquefaction-induced settlement under overburden	Saturated or dry clean sand (less effective in finer sand)	60 (ineffective above 12 ft depth)	Rapid, simple, good under water, soft underlayers may damp vibrations, hard to penetrate overlayers
Vibratory rollers	Densify by vibration, liquefaction induced settlement under roller weight	Cohesionless soils	6 to 9	Best method for thin layers or lifts
Dynamic compaction (consolidation) or heavy tamping	Repeated high-intensity impacts at the surface give immediate settlement	Cohesionless soils best, other soils can be improved	45 to 60	Simple, rapid, must protect from personal injury and property damage from flying debris; groundwater must be > 6 ft below surface, faster than preloading but less uniform
Vibroflotation	Densify by horizontal vibration and compaction of backfill material	Cohesionless soil with less than 20 percent fines	90	Economical and effective in saturated and partly saturated granular soils
Hydrocompaction	Densify by vibration or repeated impact on surface of prewetted soil	Collapsible soil	< 10	Most effective method to densify silty loose collapsible sands
b. Compaction Piles				
Compaction piles	Densify by displacement of pile volume and vibration during driving	Loose sandy soils, partly saturated clayey soils, loess	60 (limited improvement above 3 to 6)	Useful in soils with fines, uniform compaction, easy-to-check results, slow
Sand compaction piles	Sand placed in driven pipe; pipe partially withdrawn and redriven using vibratory hammer	All	——	Compressed air may be used to keep hole open as casing is partially withdrawn

Table 6-1. Continued

Method (1)	Principle (2)	Most suitable soils and types (3)	Maximum effective treatment, depth (ft) (4)	Advantages and limitations (5)
c. Precompression				
Preloading	Load applied sufficiently in advance of construction to precompress soil	Normally consolidated soft clays, silts, organic deposits, landfills	——	Easy, uniform, long time required (use sand drains or strip drains to reduce time)
Surcharge fills	Fill exceeding that required to achieve a given settlement, shorter time, excess fill removed	Same as for preloading	——	Faster than preloading without surcharge (use sand or strip drains to reduce time)
Electro-osmosis	Direct current causes water flow from anode toward cathode where it is removed	Normally consolidated silts and clays	30 to 60	No fill loading required, use in confined areas, fast, nonuniform properties between electrodes, useless in highly pervious soil
d. Reinforcement				
Mix-in-place piles and walls	Lime, cement, or asphalt placed by rotating auger or in place mixer	All soft or loose inorganic soils	> 60	Uses native soil, reduced lateral support required during excavation, difficult quality control
Strips and membranes	Horizontal tensile strips or membranes buried in soil under footings	All	< 10	Increased allowable bearing capacity; reduced deformations
Vibro-replacement stone	Hole jetted in soft fine-grain soil, and backfilled with densely compacted gravel	Very soft to firm soils (undrained strength 0.2 to 05 tsf)	60	Faster than precompression, avoids dewatering required for remove and replace, limited bearing capacity
Vibro-displacement stone	Probe displaces soil laterally, backfill discharged through probe or placed in layers after probe removed	Soft to firm soils (undrained strength 0.3 to 0.6 tsf)	50	Best in low-sensitivity soils with low groundwater
e. Grouting and Injection				
Particulate grouting	Penetration grout fills soil voids	Medium to coarse sand and gravel	Unlimited	Low cost, grout high strength
Chemical grouting	Solutions of two or more chemicals react in soil pores to form gel or soil precipitate	Medium silts and coarser	Unlimited	Low viscosity, controllable gel time, good water shutoff, high cost, hard to evalute

Table 6-1. Concluded

Method (1)	Principle (2)	Most suitable soils and types (3)	Maximum effective treatment, depth (ft) (4)	Advantages and limitations (5)
		e. Grouting and Injection		
Pressure-injected lime and lime–fly ash	Lime slurry and lime–fly ash slurry injected to shallow depths under pressure	Expansive clays, silts, and loose sands	Unlimited (usually 6 to 9)	Rapid and economical for foundations under light structures, fly ash with lime may increase cementation and strength and reduce permeability
Displacement or compaction grout	Highly viscous grout acts as radial hydraulic jack when pumped under high pressure	Soft fine-grained soils, soils with large voids or cavities	40	Corrects differential settlement, fills large voids, requires careful control
Jet grouting	Cement grouts injected to replace and mix with soils eroded by high-pressure water jet ("soilcrete" column)	Alluvial, cohesive, sandy, gravelly soils; miscellaneous fill, and others	Unlimited	Increases soil strength and decreases permeability, wide application
Electrokinetic injection	Stabilizing chemicals moved into soil by electro-osmosis	Saturated silts, silty clays	unknown	Soil and structure not subject to high pressures, useless in pervious soil
		f. Miscellaneous		
Remove and replace	Soil excavated, replaced with competent material, or improved by drying or admixture and recompacted	Inorganic soil	< 30	Uniform, controlled when replaced, may require large area dewatering
Moisture barriers	Water access to foundation soil is minimized and more uniform	Expansive soil	15	Best for small structures and pavements, may not be 100 percent effective
Prewetting	Soil is brought to estimated final water content prior to contruction	Expansive soil	6	Low cost, best for small light structures, soil may still shrink and swell
Structural fills	Structural fill distributes loads to underlying soft soils	Soft clays or organic soils, marsh deposits	——	High strength, good load distribution to underlying soft soils

Table 6-2. Range of Particle Sizes for Various Soil Improvement Methods

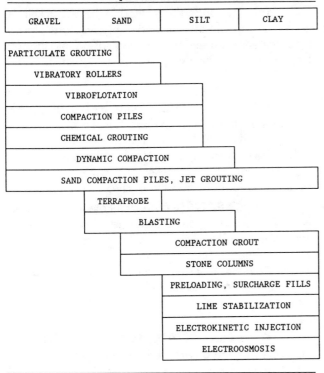

GRAVEL	SAND	SILT	CLAY

consist of a filter fabric sleeve or jacket made of non-woven polyester or polypropylene surrounding a plastic core. The drain is inserted into the soil using an installation mast containing a hollow mandrel or lance through which the drain is threaded. An anchor plate is attached to the end of the drain. Theoretical estimates of the rate of settlement are largely qualitative unless prior experience is available from similar sites because the analysis is sensitive to soil input parameters, particularly the coefficient of consolidation and existence of pervious bands of soil. Strip drains have largely replaced sand drains in practice.

(3) Stone or Chemically Stabilized Soil Columns. Columns made of stone or chemically stabilized soil increase the stiffness of the foundation and can substantially decrease settlement. Columns may fail by bulging if the adjacent soil gives inadequate support or fail by shear as a pile because of insufficient skin friction and end bearing resistance.

(a) Stone columns are made by vibro-replacement (wet) or vibro-displacement (dry) methods (Table 6-1). Diameters range from 1.5 to 4 ft with spacings from 5 to 12 ft. A blanket of sand and gravel or a semirigid mat of reinforced earth is usually placed over stone column reinforced soil to improve load transfer to the columns by arching over the in situ soil. Stone columns are not recommended for soils with sensitivities greater than 5.

(b) Lime columns are made by mixing metered or known amounts of quicklime using drilling rigs to achieve concentrations of 5 to 10 percent lime by weight of dry soil. Structures are constructed on thin concrete slabs where settlement is assumed uniform over the entire area.

(c) Cement columns are made by adding 10 to 20 percent cement as a slurry. These columns are brittle, have low permeability, and have been used below sea level.

(4) Jet Grouting. Jet grouting is the controlled injection of cement grouts to replace almost any type of soil; this soil is eroded by water jets while grouting. The most common application has been underpinning of existing structures to reduce total and differential settlement and as cutoff walls for tunnels, open cuts, canals, and dams. Jet grouting may also be used to consolidate soft foundation soils for new structures, embankments, and retaining walls. Other applications include support of excavations for open cuts and shafts and slope stabilization.

(a) Jet grouting can either break up the soil and mix grout with the natural soil particles or break up the soil, partially remove the soil, and mix grout with the remaining soil particles.

(b) Jet grouting can substantially increase the strength and stiffness of soft clay soil to reduce settlement and substantially reduce the permeability of sandy soil.

(c) Jet grouting is generally used with rapid-set cement and fly ash. Fly ash when mixed with cement or lime produces a cementatious material with excellent structural properties. Other chemicals may be used instead of cement.

(d) A single jet nozzle can be used to both break down the soil structure and force mixing of grout with the natural soil. A water jet can also be sheathed in a stream of compressed air to erode the soil while a grout jet beneath the water jet replaces the broken or disturbed soil. Diameter and discharge pressure of the nozzles, withdrawal and rotation rates, type and quality of grout, and soil type influence volume and quality of the grouted mass. Withdrawal rates and nozzle pressures are the primary design factors. Withdrawal rates vary from 1 to 50 in./min and nozzle pressures often range from 3000 to 9000 psi depending on the type of soil.

(5) Dynamic Compaction (Consolidation). Weights from 5 to 40 tons and more may be dropped from heights of 20 to 100 ft following a particular pattern for each site. The impact appears to cause partial liquefaction of granular deposits, thereby allowing the soil to settle into a more dense state.

(6) Removal by Displacement. Sufficient cohesionless fill is placed to cause bearing failures in the underlying soft soil. The soft soil is displaced in the direction of least resistance, which is usually ahead of the em-bankment fill. The displaced soil causes a mudwave that should be excavated at the same rate that the embankment is placed to minimize trapping pockets of soft soil beneath the embankment.

(7) Lightweight Fills. Sawdust, expanded foam plastic blocks, expanded shale or clay, oyster shell, and fly ash fills can partially replace excavated heavier soft material and reduce the net increase in pressure on underlying soft soil. The availability of lightweight fill in sufficient quantity at reasonable cost and suitable locations to dispose of the excavated soft soil limit application of this method.

(8) Structural (Self-Supporting Fills). Some naturally occurring materials such as dead oyster shell can form a barge-like structure from particle interlocking. Fills of loose shell have been used for highway embankments and foundations for flexible facilities such as warehouses on marsh and swamp deposits.

(9) Blasting. Cohesionless saturated sands (less than 25 percent passing the 200 mesh) are most responsive to densification by the detonation of dynamite charges in loose deposits. Soft soils that can be liquefied or displaced by advancing fill can be removed by blasting for embankment construction. Soft soils may be displaced by blasting or toe shooting in

front of the embankment. The extent of soil improvement by blasting is often uncertain.

(a) The underfill method, where backfill is placed on top of soft soil and explosives are placed under the embankment by lowering casing into the soft deposits, is most effective when the embankment width is less than 60 ft.

(b) The ditching method, where fill is placed immediately into excavations made by blasting, is effective for depths of soft soil less than 15 ft.

(c) The relief method may be useful where ditches are blasted along each side of the embankment to provide lateral stress relief and force soft shallow soil to move laterally into the ditches.

B. EXPANSIVE SOIL. Potentially expansive soils are usually desiccated and will absorb available moisture. These soils can be made to maintain volume changes within acceptable limits by controlling the soil water content and reducing the potential of the soil to heave. Methods for improving the performance of foundations in expansive soil are illustrated in Table 6-3.

C. COLLAPSIBLE SOIL. Collapsible soils settle when wetted or vibrated; therefore, the usual approach toward optimizing performance of structures on collapsible soil is prewetting the construction site. Hydrocompaction (see Table 6-1) of the site prior to

Table 6-3. Improving Performance in Expansive Soil

Method (1)	Description (2)
Removal by excavation and replacement with nonexpansive fill	Removal of surface expansive soil to depths of 4 to 8 ft and replacement with compacted nonexpansive fill usually eliminates most potential soil heave because the depth of moisture change is often limited to about 8 ft.
Placement of vertical moisture barriers	Vertical moisture barriers placed adjacent to pavements or around the perimeter of foundations down to the maximum depth of mositure changes is effective in maintaining uniform soil moisture within the barrier. Differential movements are minimized. Long-term soil wetting with uniform heave beneath impervious foundations may occur from lack of natural evapotranspiration.
Lime stabilization	Lime injected or mixed into expansive soil can reduce potential for heave by reducing the mass permeability, thereby reducing the amount of water seeping into the soil, by cementation and exchange of sodium for calcium ions. Fissures should exist in situ to promote penetration of lime-injected slurry. Lime may be detrimental in soils containing sulfates.
Potassium injection	Postassium solutions injected into expansive soil can cause a base exchange, increase the soil permeability, and effectively reduce the potential for swell.
Prewetting	Free water is added by ponding to bring soil to the estimated final water content prior to construction. Vertical sand drains may promote wetting of subsurface soil.
Surcharge	Placing 1 or 2 ft or more of permanent compacted fill on the surface of a level site prior to construction increases the overburden pressure on the underlying soil, reducing the negative (suction) pore water pressure; therefore, the potential for swell is less and tends to be more uniform. This fill also increases elevation of the site providing positive drainage of water away from the structure.

construction is commonly recommended. Chemical stabilization with lime, sodium silicate, or other chemicals is not always successful. Methods applicable to improving performance of structures on collapsible soil are illustrated in Table 6-4.

6-3. Foundation Techniques

Foundation design and construction methods can minimize soil volume changes and differential movement.

A. FLOATING FOUNDATIONS. Foundation elements such as mats and footings can be placed in excavations of sufficient depth where the pressure applied by the structure to the underlying foundation soil approximately balances the pressure applied by the excavated soil. Observed deformation will be elastic recompression settlement. The exposed soil in the bottom of the excavation must be protected from disturbance and deterioration.

B. RIBBED MATS. Slab foundations supported by a grid of stiffening beams can transfer structural loads to soil of adequate stiffness and bearing capacity. The stiffness of ribbed mats also reduces differential movement in expansive soil. The depth of stiffening beams normally does not exceed 3 ft. Ribbed mats supported on compacted cohesive nonexpansive fills are commonly constructed in expansive soil areas.

C. LEVELING JACKS. Structures may be supported by jacks on isolated footings in which the elevation can be periodically adjusted to reduce distress from excessive differential movement. Proper adjustment of leveling jacks requires periodic level surveys to determine the amount and direction of adjustment, whether up or down, and frequency of adjustment to minimize differential movement. Leveling

jacks are usually inconvenient to owner/operators of the structure.

D. DEEP FOUNDATIONS. Structural loads can be transferred to deep, firm bearing strata by piles or drilled shafts to eliminate or minimize effects of shallow soil movements on structural performance. Uplift thrust from skin friction on the perimeter of deep foundation piles or drilled shafts in expansive soil or downdrag in consolidating or collapsing soil should be considered in the design. Refer to TM 5-809-7, Design of Deep Foundations, for further details.

E. CONSTRUCTION AIDS FOR EXCAVATIONS. Settlement or loss of ground adjacent to excavations may become excessive. Causes of loss of ground include lateral rebound of perimeter walls into the excavation, rebound at the bottom of the excavation, and dewatering. Damage may occur in adjacent structures including pavements and utilities if loss of ground exceeds 0.5 in. or lateral movement of perimeter walls into an excavation exceeds 2 in. Level readings should be taken periodically to monitor elevation changes so that steps may be taken to avoid any damage. Construction aids include placement of bracing or retaining walls, placement of foundation loads as quickly as possible after the excavation is made, avoidance of ponding of water within excavations, and ground freezing. Load-bearing soils at the bottom of the excavation must be protected from deterioration and water content changes following exposure to the environment. Ground freezing provides temporary support and groundwater control in difficult soils and it is adaptable to most sizes, shapes, or depths of excavations. Ground freezing is accomplished by circulating a coolant, usually calcium chloride brine, through refrigeration pipes embedded in the soil. Refer to TM 5-818-5/AFM 88-5, chapter 6, for details on dewatering and groundwater control.

6-4. Flexible Techniques

Structures may be made flexible to tolerate differential movement by placing construction joints in the superstructure or by using flexible construction materials. Steel or wood frames, metal siding, wood paneling, and asphalt floors can tolerate large differential settlements or angular distortions up to about 1/150.

Section II. Remedial Methods

6-5. General

Remedial work for damaged structures is often aggravated because it is difficult to determine the

Table 6-4. Improving Performance of Collapsible Soil

Depth of soil treatment (ft) (1)	Description (2)
0 to 5	Wetting, mixing, and compaction
> 5	Overexcavation and recompaction with or without chemical additives such as lime or cement
> 5	Hydrocompaction
> 5	Vibro-flotation
> 5	Lime pressure injection
> 5	Sodium silicate injection
> 5	Prewetting by ponding, vertical sand drains promote wetting of subsurface soil

cause of the problem (e.g., location of source or loss of soil moisture with swelling or settling of expansive/collapsible soil may not be readily apparent). Investigation and repair are specialized procedures that usually require much expertise and experience. Cost of repair work can easily exceed the original cost of the foundation. Repair of structures in heaving soil is usually much more costly than in settling soil. Structures are less able to tolerate the tensile strains from heaving soil than the compressive strains in settling soil. The amount of damage that requires repair also depends on the attitudes of the owner and effected people to tolerate distortion and consequences if the distortion and damage are ignored. Only one remedial procedure should be attempted at a time after a course of action has been decided so that its effect on the structure may be determined. Several common remedial methods are discussed here. Refer to TM 5-818-7, Foundations in Expansive Soils, for further details on remedial methods for foundations.

6-6. Underpinning with Piles

Underpinning may be accomplished by a variety of methods: drilled-in-place tangent piles, cast-in-place rigid concrete slurry walls, precast concrete retaining walls, root or pin piles, concrete underpinning pits, and jacked steel piles. Selection of the underpinning method depends on the nature of the subgrade soil and its expected behavior during underpinning. Refer to TM 5-809-7, Deep Foundations, for details on piles.

A. AVOID GROUND LOSS. Possibility of ground loss during installation may eliminate use of tangent piles, slurry walls, and precast concrete retaining walls.

B. INTERFERENCE WITH UTILITIES. Underground utilities may eliminate use of piles or cast-in-place concrete shafts.

6-7. Grouting

Structures may be stabilized by injecting Portland cement, fine soil, and chemicals into the problem soil. Grouting mixtures usually consist of fine soil, Portland cement, and water; lime and water; sodium silicate; calcium chloride; polymers; and resins. Jet and compaction grouting, for example, reduce differential settlement of structures. Compaction grouting can raise a structure that has settled. The stiffness and strength of the soil may be increased by injecting a grout containing additives such as Portland cement to improve the performance of the soil. Compaction grouting may use 12 to 15 percent by weight of Portland cement mixed with soil and water to make a viscous low-slump grout that is to be pumped into bored holes at pressures up to 500 psi. Refer to TM 5-818-6, Grouting Methods and Equipment; EM 1110-2-3504, Chemical Grouting; and EM 1110-2-3506, Grouting Technology, for details on grouting.

6-8. Slabjacking

Slabjacking, the lifting or leveling of distorted foundations, is usually faster than other solutions for remedial work. Grouting materials include Portland cement, hydrated lime, fly ash, asphalt bitumen, drilling mud, casting plaster, and limestone dust. Consistency of the grout varies from less commonly used thin fluids to more common heavy pourable or stiff mortar-like mixtures (with nearly zero slump). Cement contents vary from 3 to 33 percent with sand or soil materials all passing the No. 16 sieve. Leakage from joints and along the edges of slabs can present serious problems, which are commonly offset by increasing the consistency of the grout. Lifts of as much as 1 ft are common. Properly performed slabjacking will not usually cause new fractures in the foundation, but existing cracks tend to open. Experience is required to cause low points to rise while maintaining high points at a constant elevation.

APPENDIX A

REFERENCES

1. TM 5-809-1/AFM 88-3, Chapter 1, "Load Assumptions for Buildings."

2. TM 5-809-7, "Design of Deep Foundations (Except Hydraulic Structures.)"

3. TM 5-818-1/AFM 88-5, Chapter 7, "Procedures for Foundation Design of Buildings and Other Structures (Except Hydraulic Structures)."

4. TM 5-818-5/AFM 88-5, Chapter 6, NAVFAC P-418, "Dewatering and Groundwater Control."

5. TM 5-818-6, "Grouting Methods and Equipment."

6. TM 5-818-7, "Foundations in Expansive Soils."

7. ER 1110-2-1806, "Earthquake Design and Analysis for Corps of Engineers Dams."

8. EM 1110-1-1804, "Geotechnical Investigations."

9. EM 1110-2-1903, "Bearing Capacity."

10. EM 1110-2-1906, "Laboratory Soils Testing."

11. EM 1110-2-1907, "Soil Sampling."

12. EM 1110-2-1911, "Construction Control for Earth and Rock-Fill Dams."

13. EM 1110-2-1913, "Design and Construction of Levees."

14. EM 1110-2-2102, "Waterstop and Other Joint Materials."

15. EM 1110-2-2300, "Earth and Rockfill Dams, General Design and Construction Considerations."

16. EM 1110-2-3506, "Grouting Technology."

17. EM 1110-2-3504, "Chemical Grouting."

18. Department of the Navy, NAVFAC DM-7.1, May 1982, "Soil Mechanics." Available from Naval Facilities Engineering Command, 200 Stovall Street, Alexandria, VA 22332.

19. Department of the Navy, NAVFAC DM-7.3, April 1983, "Soil Dynamics." Available from Naval Facilities Engineering Command, 200 Stovall Street, Alexandria, VA 22332.

20. Nuclear Regulatory Commission, Regulatory Guide 1.6, "Design Response Spectra for Seismic Design of Nuclear Power Plants." Available from Nuclear Regulatory Commission, 1717 H Street, Washington, DC 20555.

21. American Society for Testing and Materials Standard Methods of Test D 1194, "Bearing Capacity of Soil for Static Load on Spread Footings." Available from American Society for Testing and Materials, 1916 Race Street, Philadelphia, PA 19103.

22. American Society for Testing and Materials Standard Methods D 1557, "Moisture-Density Relations of Soils and Soil-Aggregate Mixtures Using 10-lb (4.54-kg) Rammer and 18-in. (457mm) Drop." Available from American Society for Testing and Materials, 1916 Race Street, Philadelphia, PA 19103.

23. American Society for Testing and Materials Standard Method D 1586, "Penetration Test and Split-Barrel Sampling of Soils." Available from American Society for Testing and Materials, 1916 Race Street, Philadelphia, PA 19103.

24. American Society for Testing and Materials Standard Methods of Test D 2435, "One-Dimensional Consolidation Properties of Soils." Available from American Society for Testing and Materials, 1916 Race Street, Philadelphia, PA 19103.

25. American Society for Testing and Materials Standard Methods of Test D 4546, "One-Dimensional Swell or Settlement Potential of Cohesive Soils." Available from American Society for Testing and Materials, 1916 Race Street, Philadelphia, PA 19103.

26. Canadian Geotechnical Society, 1985. "Canadian Foundation Engineering Manual," second edition. Available from Canadian Geotechnical Society, BiTech Publishers Ltd., 801 1030 W. Georgia Street, Vancouver, BC, V6E 2Y3 Canada.

APPENDIX B

BIBLIOGRAPHY

1. Alpan, I. 1964. "Estimating the Settlements of Foundations on Sands," *Civil Engineering and Public Works Review*, Vol. 59, pp. 1415–1418. Available from Morgan-Grampian Ltd., 30 Calderwood Street, Woolwich, London SE18 GQH, England.

2. Baladi, G. 1968. "Distribution of Stresses and Displacements Within and Under Long, Elastic and Viscoelastic Embankments." Available from Purdue University, West Lafayette, IN 47907.

3. Brown, P.T. 1969. "Numerical Analysis of Uniformly Loaded Circular Rafts on Deep Elastic Foundations," *Geotechnique*, Vol. 19, pp. 399–404, the Institution of Civil Engineers. Available from Thomas Telford Ltd., 1-7 Great George Street, Westminster, London SW1P 3AA, England.

4. Burland, J.B., and Burbidge, M.C. 1985. "Settlement of Foundations on Sand and Gravel," *Proceedings, Institution of Civil Engineers, Part 1*, Vol. 78, pp. 1325–1381. Available from Thomas Telford Ltd., 1-7 Great George Street, Westminster, London SW1P 3AA, England.

5. Burmister, D.M. 1954. "Influence Diagram of Stresses and Displacements in a Two-Layer Soil System with a Rigid Base at a Depth H," Contract No. DA-49-129-ENG-171 with U.S. Army Corps of Engineers, Columbia University, New York, NY. Available from National Technical Information Service, 5285 Port Royal Road, Springfield, VA 22161.

6. Burmister, D.M. 1963. "Physical, Stress-Strain, and Strength Responses of Granular Soils," *Field Testing of Soils*, ASTM Special Technical Publication No. 322, pp. 67–97. Available from American Society for Testing and Materials, 1916 Race Street, Philadelphia, PA 19103.

7. Burmister, D.M. 1965. "Influence Diagrams for Stresses and Displacements in a Two-Layer Pavement System for Airfields," Department of the Navy, Washington, DC. Available from Department of the Navy, Washington, DC 20350.

8. Cargill, K.W. 1985. "Mathematical Model of the Consolidation/Desiccation Processes in Dredged Material," Technical Report D-85-4. Available from Research Library, U.S. Army Engineer Waterways Experiment Station, Vicksburg, MS 39180.

9. Christian, J.T., and Carrier, W.D., III. 1978. "Janbu, Bjerrum and Kjaernsli's Chart Reinterpreted," *Canadian Geotechnical Journal.* Available from National Research Council of Canada, Resarch Journals, Ottawa, Ontario K1A OR6, Canada.

10. Committee on Placement and Improvement of Soils. 1978. "Soil Improvement—History, Capabilities, and Outlook," and 1987, "Soil Improvement—A Ten Year Update." Available from American Society of Civil Engineers, 345 East 47th Street, New York, NY 10017.

11. Costet, J., and Sanglerat, G. 1975. "Cours Practique de Mechanique des Sols," *Plasticite et Calcul des Tassements*, Vol. 1, second edition, p. 117. Available from Dunod, 24–26 Boulevard de L'hopital 75006, Paris, France.

12. D'Appolonia, D.J., D'Appolonia, E.D., and Brissette, R.F. 1970. "Closure: Settlement of Spread Footings on Sand," *Journal of the Soil Mechanics and Foundations Division*, Vol. 96, pp. 754–762. Available from American Society of Civil Engineers, 345 East 47th Street, New York, NY 10017.

13. D'Orazio, T.B., and Duncan, J.M. 1987. "Differential Settlement in Steel Tanks," *Journal of the Geotechnical Engineering Division*, Vol. 113, pp. 967–983. Available from American Society of Civil Engineers, 345 East 47th Street, New York, NY 10017.

14. Duncan, J.M., and Chang, C.Y. 1970. "Nonlinear Analysis of Stress and Strain in Soils," *Journal of the Soil Mechanics and Foundations Division*, Vol. 96, pp. 1629–1653. Available from American Society of Civil Engineers, 345 East 47th Street, New York, NY 10017.

15. Duncan, J.M., D'Orazio, T.B., Chang, C.-S., and Wong, K.S. 1981. "CON2D: A Finite Element Computer Program for Analysis of Consolidation," Geotechnical Engineering Report No. UCB/GT/81-01, University of California, Berkeley. Available from University of California, Berkeley 94720.

16. Feda, J. 1966. "Structural Stability of Subsident Loess Soil from Prahadejuice," *Engineering Geology*, Vol. 1, pp. 201–219. Available from Elsevier Science Publishers B.V., Box 211, 1000 AE, Amsterdam, The Netherlands.

17. Feld, J. 1965. "Tolerance of Structures to Settlement," *Journal of the Soil Mechanics and Foundations Division*, Vol. 91, pp. 63–77. Available from American Society of Civil Engineers, 345 East 47th Street, New York, NY 10017.

18. Gibbs, H.J., and Bara, J.P. 1962. "Predicting Surface Subsidence from Basic Soil Tests," *Field Testing of Soils*, Special Technical Publication No. 322, pp. 231–247. Available from American Society for Testing and Materials, 1916 Race Street, Philadelphia, PA 19103.

19. Gibson, R. E. 1967. "Some Results Concerning Displacements and Stresses in a Nonhomogeneous Elastic Half-Space," *Geotechnique*, Vol. 17, pp. 58–67. Available from Thomas Telford Ltd., 1-7 Great George Street, Westminster, London SW1P 3AA, England.

20. Gogoll, F.H. 1970. "Foundations in Swelling Clay Beneath a Granular Blanket," *Proceedings of the Symposium on Soils and Earth Structures in Arid Climates*, Adelaide, Australia, pp. 42–48. Available from Institution of Engineers, Miadna Pty. Ltd., Box 588, Crons Nest N5W 2065, Australia.

21. Harr, M.E. 1966. *Foundations of Theoretical Soil Mechanics*, pp. 98–99. Available from McGraw-Hill, 1221 Avenue of the Americas, New York, NY 10020.

22. Houston, S.L., Houston, W.N., and Spadola, D.J. 1988. "Prediction of Field Collapse of Soils due to Wetting," *Journal of Geotechnical Engineering*, Vol. 114, pp. 40–58. Available from American Society of Civil Engineers, 345 East 47th Street, New York, NY 10017.

23. Hughes, J.M.O. 1982. "Interpretation of Pressuremeter Tests for the Determination of Elastic Shear Modulus," *Proceedings of Engineering Foundation Conference on Updating Subsurface Sampling of Soils and Rocks and Their In-Situ Testing*, pp. 279–289, Santa Barbara, CA. Available from American Society of Civil Engineers, 345 East 47th Street, New York, NY 10017.

24. Hyde, A.F., and Brown, S.F. 1976. "The Plastic Deformation of a Silty Clay Under Creep and Repeated Loading," *Geotechnique*, Vol. 26, pp. 173–184, The Institution of Civil Engineers. Available from Thomas Telford Ltd., 1–7 Great George Street, Westminster, London SW1P 3AA, England.

25. Janbu, N., and Senneset, K. 1981. "Settlement due to Drained, Cyclic Loads," *Proceedings of the 10th International Conference on Soil Mechanics and Foundation Engineering*, Vol. 1, pp. 165–170, Stockholm, Sweden. Available from A.A. Balkema, Box 1675, Rotterdam, The Netherlands.

26. Jennings, J.E., and Knight, K. 1975. "A Guide to Construction on or with Materials Exhibiting Additional Settlement due to 'Collapse' of Grain Structure," *Proceedings of the Sixth Regional Conference for Africa on Soil Mechanics and Foundation Engineering*, pp. 99–105, Durban. Available from A.A. Balkema, Box 1675, Rotterdam, The Netherlands.

27. Jeyapalan, J., and Boehm, R. 1986. "Procedures for Predicting Settlement in Sands," *Settlement of Shallow Foundations on Cohesionless Soils: Design and Performance*, Geotechnical Special Publication No. 5, pp. 1–22. Available from American Society of Civil Engineers, 345 East 47th Street, New York, NY 10017.

28. Johnson, L.D. 1989. "Design and Construction of Mat Foundations," Miscellaneous Paper GL-89-27. Available from Research Library, U.S. Army Engineer Waterways Experiment Station, Vicksburg, MS 39180.

29. Johnson, L.D. 1989. "Performance of a Large Ribbed Mat on Cohesive Soil," *Foundation Engineering: Current Principles and Practices*, Vol. 1, F.H. Kulhawy, Ed. Available from American Society of Civil Engineers, 345 East 47th Street, New York, NY 10017.

30. Johnson, L.D., and Snethen, D.R. 1979. "Prediction of Potential Heave of Swelling Soils," *Geotechnical Testing Journal*, Vol. 1, pp. 117–124. Available from American Society for Testing and Materials, 1916 Race Street, Philadelphia, PA 19103.

31. Kay, J.N., and Cavagnaro, R.L. 1983. "Settlement of Raft Foundations," *Journal of the Geotechnical Engineering Division*, Vol. 109, pp. 1367–1382. Available from American Society of Civil Engineers, 345 East 47th Street, New York, NY 10017.

32. Komornik, A., and David, D. 1969. "Prediction of Swelling Pressure of Clays," *Journal of the Soil Mechanics and Foundations Division*, Vol. 95, pp. 209–225. Available from American Society of Civil Engineers, 345 East 47th Street, New York, NY 10017.

33. Krinitzsky, E.L., and Chang, F.K. 1987. "Parameters for Specifying Intensity-Related Earthquake Ground Motion," Report 25 of the series "State-of-the-Art for Assessing Earthquake Hazards in the United States," Miscellaneous Paper S-73-1. Available from Research Library, U.S. Army En-

gineer Waterways Experiment Station, Vicksburg, MS 39180.

34. Krinitzsky, E.L., and Chang, F.K. 1987. "Parameters for Specifying Magnitude-Related Earthquake Ground Motions," Report 26 of the series "State-of-the-Art for Assessing Earthquake Hazards in the United States," Miscellaneous Paper S-73-1. Available from Research Library, U.S. Army Engineer Waterways Experiment Station, Vicksburg, MS 39180.

35. Lambe, T.W., and Whitman, R.V. 1969. *Soil Mechanics*, pp. 202, 216. Available from John Wiley and Sons Ltd., 605 Third Ave., New York, NY 10016.

36. Leonards, G.A. 1976. "Estimating Consolidation Settlements of Shallow Foundations on Overconsolidated Clay," Special Report 163. Available from Transportation Research Board, National Research Council, Washington, DC 20418.

37. Leonards, G.A., and Frost, J.D. 1988. "Settlement of Shallow Foundations on Granular Soils," *Journal of Geotechnical Engineering*, Vol. 114, pp. 791–809. Available from American Society of Civil Engineers, 345 East 47th Street, New York, NY 10017.

38. Lowe, J., Jonas, E., and Obrician, V. 1969. "Controlled Gradient Consolidation Test," *Journal of the Soil Mechanics and Foundations Division*, Vol. 95, pp. 77–97. Available from American Society of Civil Engineers, 345 East 47th Street, New York, NY 10017.

39. Lutenegger, A.J. 1988. "Current Status of the Marchetti Dilatometer Test," *Penetration Testing 1988 ISOPT-1*, Vol. 1, pp. 137–155, Orlando, FL, J. DeRuiter, Ed. Available from A.A. Balkema, Box 1675, Rotterdam, The Netherlands.

40. Lysmer, J., and Duncan, J.M. 1969. "Stress and Deflections in Foundations and Pavements," Berkeley, CA. Available from Department of Civil Engineering, Institute of Transportation and Traffic Engineering, University of California, Berkeley, 94720.

41. Marchetti, S. 1985. "On the Field Determination of Ko in Sands," Discussion Session No. 2A, *XI International Conference on Soil Mechanics and Foundation Engineering*, San Francisco, CA. Available from A.A. Balkema, Box 1675, Rotterdam, The Netherlands.

42. Marcuson, W.F., III. 1977. "Determination of In Situ Density of Sands," *Liquefaction Potential of Dams and Foundations*, Research Report S-76-2, No. 4. Available from Research Library, U.S. Army Engineer Waterways Experiment Station, Vicksburg, MS 39180.

43. Mesri, G., and Godlewski, P. 1977. "Time- and Stress-Compressibility Interrelationships," *Journal of the Geotechnical Engineering Division*, Vol. 103, pp. 417–430. Available from American Society of Civil Engineers, 345 East 47th Street, New York, NY 10017.

44. Mitchell, J.K., and Gardner, W.S. 1975. "In Situ Measurement of Volume Change Characteristics," SOA Report, *Proceedings of the American Society of Civil Engineers Special Conference on the In Situ Measurement of Soil Properties*, Raleigh, NC. Available from American Society of Civil Engineers, 345 East 47th Street, New York, NY 10017.

45. Mosher, R.L., and Radhakrishnan, N. 1980. "Computer Programs for Settlement Analysis," Instruction Report K-80-5. Available from Research Library, U.S. Army Engineer Waterways Experiment Station, Vicksburg, MS 39180.

46. Nayak, N.V., and Christensen, R.W. 1971. "Swelling Characteristics of Compacted Expansive Soil," *Clays and Clay Minerals*, Vol. 19, pp. 251–261. Available from Allen Press Inc., 1041 New Hampshire Street, Box 368, Lawrence, KS 66044.

47. Newland, P.L., and Allely, B.H. 1960. "A Study of the Consolidation Characteristics of a Clay," *Geotechnique*, Vol. 10, pp. 62–74, The Institution of Civil Engineers. Available from Thomas Telford Ltd., 1-7 Great George Street, Westminster, London SW1P 3AA, England.

48. Northey, R.D. 1969. "Collapsing Soils: State of the Art," *Seventh International Conference of Soil Mechanics and Foundation Engineering*, Vol. 5, pp. 445. Available from Sociedad Mexicana de Mecanica de Suelos, A.C., Mexico City, Mexico.

49. O'Neill, M.W., and Ghazzaly, O.I. 1977. "Swell Potential Related to Building Performance," *Journal of Geotechnical Engineering Division*, Vol. 103, pp. 1363–1379. Available from American Society of Civil Engineers, 345 East 47th Street, New York, NY 10017.

50. Peck, R.B., and Bazarra, A. 1969. "Discussion of Settlement of Spread Footings on Sand," by D'Appolonia et al., *Journal of the Soil Mechanics and Foundations Division*, Vol. 95, pp. 905–909. Available from American Society of Civil Engineers, 345 East 47th Street, New York, NY 10017.

51. Peck, R.B., Hanson, W.F., and Thornburn, T.H. 1974. *Foundation Engineering*, second edition, pp. 307–314. Available from John Wiley and Sons, 605 Third Ave., New York, NY 10016.

52. Perloff, W. H. 1975. "Pressure Distribution and Settlement," Chapter 4, *Foundation Engineering*

Handbook, pp. 148–196, H.F. Winterkorn and H.Y. Fang, Eds. Available from Van Nostrand Reinhold Co., New York, NY 10020.

53. Polshin, D.E., and Tokar, R.A. 1957. "Maximum Allowable Nonuniform Settlement of Structures," *Proceedings of the Fourth International Conference on Soil Mechanics and Foundation Engineering*, Vol. 1, pp. 402–405. Available from Butterworths Publications Ltd.,88 Kingsway, London WC2, England.

54. Schmertmann, J.H. 1955. "The Undisturbed Consolidation Behavior of Clay," *Transactions*, Vol. 120, pp. 1201–1233. Available from American Society of Civil Engineers, 345 East 47th Street, New York, NY 10017.

55. Schmertmann, J.H. 1970. "Static Cone to Compute Static Settlement over Sand," *Journal of the Soil Mechanics and Foundations Division*, Vol. 96, pp. 1011–1043. Available from American Society of Civil Engineers, 345 East 47th Street, New York, NY 10017.

56. Schmertmann, J.H. 1978. *Guidelines for Cone Penetration Test Performance and Design*, Report No. FHWA-TS-78-209. Available from U.S. Department of Transportation, Federal Highway Administration, Office of Research and Development, Washington, DC 20590.

57. Schmertmann, J.H. 1983. "Revised Procedure for Calculating K_o and OCR from DMT's with I.D. > 1.2 and which Incorporate the Penetration Force Measurement to Permit Calculating the Plane Strain Friction Angle," DMT Digest No. 1. Available from GPE Inc., Gainesville, FL 32611.

58. Schmertmann, J.H. 1986. "Dilatometer to Compute Foundation Settlement," *Use of Insitu Tests in Geotechnical Engineering*, Geotechnical Special Publication No. 6, pp. 303–321. Available from American Society of Civil Engineers, 345 East 47th Street, New York, NY 10017.

59. Schmertmann, J.H., Hartman, J.P., and Brown, P.R. 1978. "Improved Strain Influence Factor Diagrams," *Journal of the Geotechnical Engineering Division*, Vol. 104, pp. 1131–1135, Available from American Society of Civil Engineers, 345 East 47th Street, New York, NY 10017.

60. Schultze, E., and Sherif, G. 1973. "Prediction of Settlements from Evaluated Settlement Observations for Sand," *Proceedings of the Eighth International Conference on Soil Mechanics and Foundation Engineering*, Vol. 1, pp. 225–230. Available from USSR National Society for Soil Mechanics and Foundation Engineers, Gosstray USSR, Marx Prospect 12, Moscow K-9.

61. Templeton, A.E. 1984. "User's Guide: Computer Program for Determining Induced Stresses and Consolidation Settlements," Instruction Report K-84-7. Available from Research Library, U.S. Army Engineer Waterways Experiment Station, Vicksburg, MS 39180.

62. Tokimatsu, K., and Seed, H.B. 1984. "Simplified Procedures for the Evaluation of Settlements in Sands due to Earthquake Shaking," Report No. UCB/EERC-84/16. Available from Earthquake Engineering Research Center, University of California, Berkeley, 94720.

63. Tokimatsu, K., and Seed, H.B. 1987. "Evaluation of Settlements in Sands due to Earthquake Shaking," *Journal of Geotechnical Engineering*, Vol. 113, pp. 861–878. Available from American Society of Civil Engineers, 345 East 47th Street, New York, NY 10017.

64. Vijayvergiya, V.N., and Ghazzaly, O.I. 1973, "Prediction of Swelling Potential for Natural Clays," *Proceedings of the Third International Conference on Expansive Clay Soils*, Vol. 1. Available from Jerusalem Academic Press, Jerusalem.

65. Wahl, H.E. 1981. "Tolerable Settlement of Buildings," *Journal of the Geotechnical Engineering Division*, Vol. 107, pp. 1489–1504. Available from American Society of Civil Engineers, 345 East 47th Street, New York, NY 10017.

66. Winterkorn, H.F., and Fang, H.-Y. 1975. "Soil Technology and Engineering Properties of Soils," Chapter 2, *Foundation Engineering Handbook*, H.F. Winterkorn and H-Y Fang, Eds., pp. 112–117. Available from Van Nostrand Reinhold Co., 135 West 50th Street, New York, NY 10020.

67. Wissa, A.E.Z., Christian, J.T., and Davis, E.H. 1971. "Consolidation at Constant Rate of Strain," *Journal of the Soil Mechanics and Foundations Division*, Vol. 97, pp. 1393–1414. Available from American Society of Civil Engineers, 345 East 47th Street, New York, NY 10017.

APPENDIX C

STRESS DISTRIBUTION IN SOIL

C-1. General

Displacements in soil occur from stresses applied by loading forces. The distribution of stress in soil should be known for realistic estimates of displacements caused by applied loads on the supporting soil.

A. EFFECT OF FOUNDATION STIFF-NESS. The distribution of stress in soil depends on the contact pressure between the foundation and soil, which is a function of the relative stiffness K_R between the soil and the foundation (item 3)

$$K_R = \frac{E_f(1 - v_s^2)}{E_s}\left[\frac{D_f}{R}\right]^3 \qquad \text{(C-1)}$$

where E_f = Young's modulus of foundation in tsf; v_s = Poisson's ratio of foundation soil; E_s = Young's modulus of foundation soil in tsf; D_f = thickness of foundation in ft; and R = radius of foundation in ft.

A uniformly loaded flexible foundation where stiffness $K_R < 0.1$ causes a uniform contact pressure, whereas a uniformly loaded rigid foundation where $K_R > 10$ causes a highly nonuniform contact pressure distribution (Figure 1-3).

(1) Embankments. Earth embankments are flexible and normally in full contact with the supporting soil.

(2) Foundations for Structures. Foundations such as large mats and footings with sufficient stiffness ($K_R > 0.1$) may not always be in complete contact with the soil.

B. LIMITING CONTACT PRESSURES. Contact pressures are limited to maximum pressures defined as the bearing capacity. Refer to EM 1110-2-1903, Bearing Capacity, for estimation of the soil-bearing capacity.

C. OTHER FACTORS INFLUENCING CONTACT PRESSURE. The distribution of contact pressures is also influenced by the magnitude of loading, depth of applied loads, size, shape, and method of load application such as static or dynamic applied loads.

C-2. Evaluation of Stress Distributions in Soil

The following methods may be used to estimate the stress distribution for an applied load Q at a point and a uniform contact pressure q applied to an area. Practical calculations of settlement are based on these estimates of stress distribution.

A. APPROXIMATE 2:1 DISTRIBU-TION. An approximate stress distribution assumes that the total applied load on the surface of the soil is distributed over an area of the same shape as the loaded area on the surface, but with dimensions that increase by an amount equal to the depth below the surface (Figure C-1). At a depth z in ft below the ground surface, the total load Q in tons applied at the ground surface by a structure is assumed to be uniformly distributed over an area $(B + z)$ by $(L + z)$. The increase in vertical pressure $\Delta\sigma_z$ in tsf at depth z for an applied load Q is given by

$$\Delta\sigma_z = \frac{Q}{(B + z) \cdot (L + z)} \qquad \text{(C-2)}$$

where B and L are the width and length of the foundation in ft, respectively. $\Delta\sigma_z$ may be the pressure σ_{st} caused by construction of the structure. Vertical stresses calculated by Equation C-2 agree reasonably well with the Boussinesq method discussed later for depths between B and $4B$ below the foundation.

B. BOUSSINESQ. The Boussinesq solution is based on the assumption of a weightless half-space free of initial stress and deformation. The modulus of elasticity is assumed constant and the principle of linear superposition is assumed valid.

(1) Equations for Strip and Area Loads. The Boussinesq solutions for increase in vertical stress $\Delta\sigma_z$ shown in Table C-1 for strip and area loads apply to elastic and nonplastic materials where deformations are continuous and unloading and reloading do not occur. Refer to EM 1110-2-1903, Bearing Capacity, and item 40 for equations of stress distributions beneath other foundation shapes.

(2) Graphical Solutions. The stress distribution beneath a corner of a rectangular uniformly

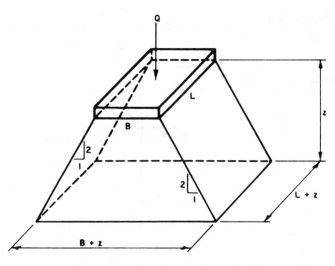

Figure C-1. Approximate stress distribution by the 2:1 method

loaded area may be evaluated from

$$\Delta\sigma_z = I_\sigma \cdot q \qquad \text{(C-3)}$$

where I_σ is the influence factor from Figure C-2 and q is the bearing pressure. Refer to TM 5-818-1 for further information on I_σ.

(a) The increase in stress beneath the center of a rectangular loaded area is given through the assumption of superposition of stresses as 4 times that given by Equation C-3.

(b) The increase in stress beneath the center of an edge is twice that given by Equation C-3.

(c) Refer to EM 1110-2-1903, Bearing Capacity, and TM 5-818-1, Procedures for Foundation Design of Buildings and Other Structures (Except Hydraulic Structures), for further details on estimation of stress distributions.

Table C-1. Boussinesq Solutions for Increase in Vertical Stress $\Delta\sigma_z$ Beneath a Foundation

Type of load normal to surface (1)	Equation for $\Delta\sigma_z$ (2)	Coordinate system (3)
a. Uniform Loads		

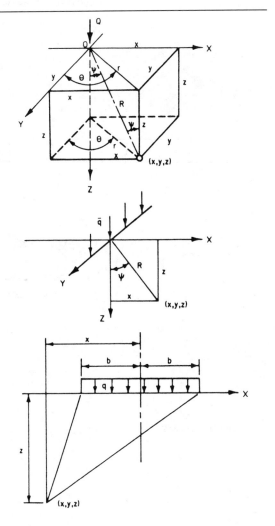

POINT

$$\frac{3\,Q}{2\,\pi\,R^2}\cos^3\psi$$

Q = NORMAL LOAD, TONS

$r^2 = x^2 + y^2$

$R^2 = r^2 + z^2$

LINE

$$\frac{2\,\bar{q}\,z^3}{\pi\,R^4}$$

$\bar{q}$ = NORMAL LOAD, TONS/FT

$R^2 = x^2 + z^2$

STRIP

$$\frac{q}{\pi}(\alpha + \sin\alpha \cos(\alpha + 2\beta))$$

q = CONTACT PRESSURE, TSF

$\alpha = \tan^{-1}\left(\frac{x+b}{z}\right) - \beta$, RADIANS

$\beta = \tan^{-1}\left(\frac{x-b}{z}\right)$, RADIANS

Table C-1. Continued

Type of load normal to surface (1)	Equation for $\Delta\sigma_z$ (2)	Coordinate system (3)
a. Uniform Loads		

AREA
(POINT UNDER
CENTER CIRCULAR
AREA)

$$q r^2 \frac{(S^2 + 2z^2)}{2S^4}$$

$$S^4 = r^2 + z^2$$

q = CONTACT PRESSURE, TSF

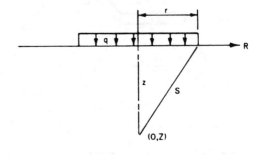

(POINT UNDER
CORNER RECTANGULAR
AREA)

$$\frac{q}{2\pi}\left[TAN^{-1}\frac{ab}{zC} + \frac{abz}{C}\left(\frac{1}{A^2} + \frac{1}{B^2}\right)\right]$$

$$A^2 = a^2 + z^2$$

$$B^2 = b^2 + z^2$$

$$C = (a^2 + b^2 + z^2)^{1/2}$$

q = CONTACT PRESSURE, TSF

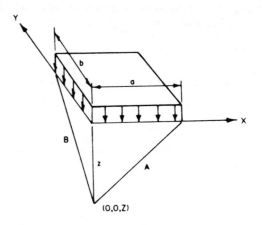

---

b. Nonuniform Strip Loads

UNIFORM ON HALF
INFINITE SPACE

$$\frac{q}{\pi}(\beta + \frac{xz}{R^2})$$

q = CONTACT PRESSURE, TSF

β = ANGLE IN RADIANS

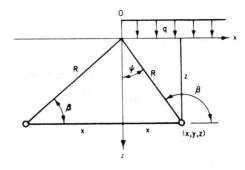

LINEAR INCREASE

$$\frac{q_0}{\pi a}(x\beta + z)$$

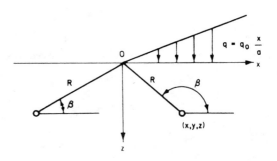

Table C-1. Concluded

Type of load normal to surface (1)	Equation for $\Delta\sigma_z$ (2)	Coordinate system (3)
b. Nonuniform Strip Loads		
TRIANGULAR	$\dfrac{q}{\pi}\left[\dfrac{x}{a}\,\alpha + \dfrac{(a+b-x)}{b}\,\beta\right]$	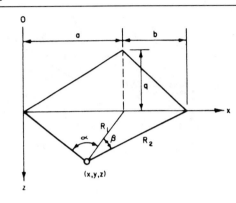
CONTRATRAPEZOIDAL	$\dfrac{q}{\pi a}\left[a\,(\beta+\beta') - b\,(\alpha+\alpha') + x\,(\alpha-\alpha')\right]$	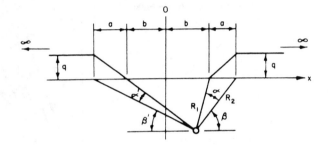
TRAPEZOIDAL	$\dfrac{q}{\pi a}\left[a(\alpha+\beta+\alpha')+b(\alpha+\alpha')+x(\alpha-\alpha')\right]$	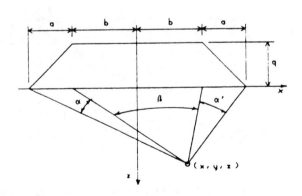

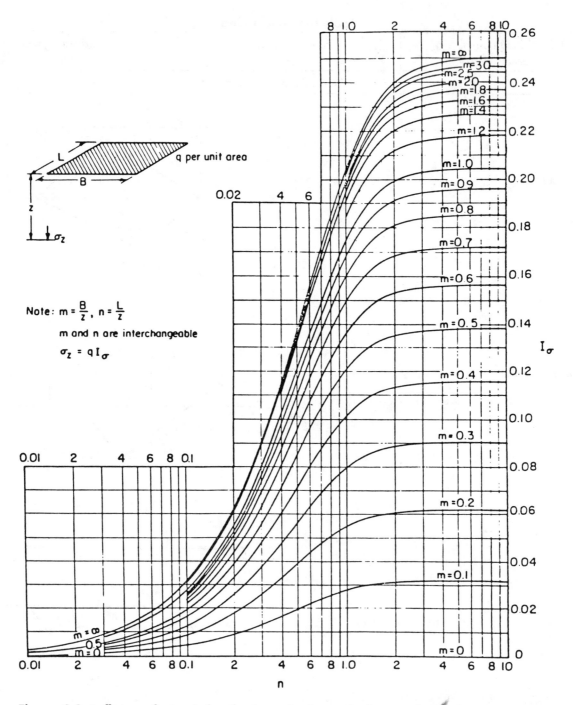

Figure C-2. Influence factor I_σ for the increase in vertical stress beneath a corner of a uniformly loaded rectangular area for the Boussinesq stress distribution (from TM 5-818-1)

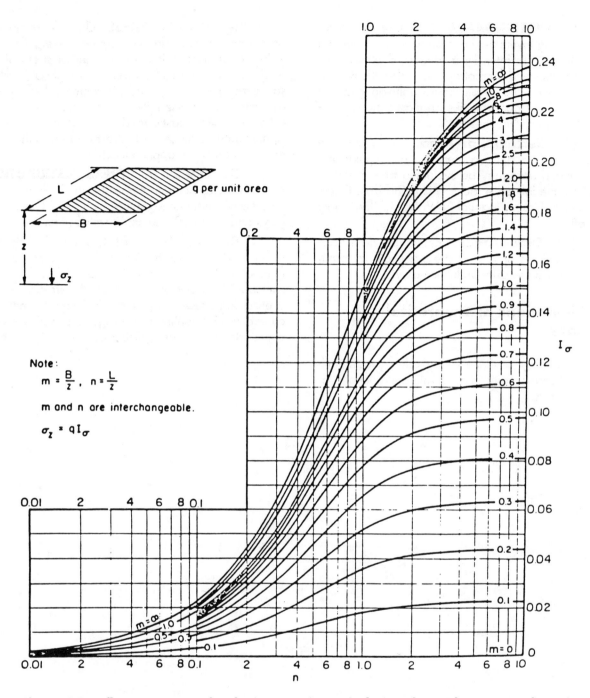

Note:
$$m = \frac{B}{z}, \quad n = \frac{L}{z}$$

m and n are interchangeable.

$$\sigma_z = q I_\sigma$$

Figure C-3. Influence Factor I_σ for the increase in vertical stress beneath a corner of a uniformly loaded rectangular area for the Westergaard stress distribution (from NAVFAC DM 7.1)

C. WESTERGAARD. Soils that are stratified with strong layers may reinforce soft layers so that the resulting stress intensity at deeper depths is less than that formulated for isotropic soil after the Boussinesq approach. The Westergaard solution assumes that stratified soil performs like an elastic medium reinforced by rigid, thin sheets.

(1) Rectangular Area. The increase in vertical stress beneath a corner of a rectangular uniformly loaded area may be evaluated from Equation C-3 where the influence factor I_σ is found from Figure C-3. Refer to Design Manual NAVFAC 7.1 for further information.

(2) Other Areas. Item 40 provides Westergaard solutions of vertical stress beneath uniformly loaded foundations using influence charts.

C-3. Limitations of Theoretical Solutions

Boundary conditions may differ substantially from idealized conditions to invalidate solutions by elasticity theory.

A. INITIAL STRESS. Elastic solutions such as the Boussinesq solution assumes a weightless material not subject to initial stress. Initial stress always exists in situ because of overburden weight of overlying soil, past stress history, and environmental effects such as desiccation. These initial stresses, through Poisson's ratio, nonlinear elastic modulus, and soil anisotropy significantly influence in situ stress and strain that occur through additional applied loads.

B. ERROR IN STRESS DISTRIBUTION. Actual stresses beneath the center of shallow footings may exceed Boussinesq values by 15 to 30 percent in clays and 20 to 30 percent in sands (item 5).

C. CRITICAL DEPTH. The critical depth z_c is the depth at which the increase in stresses $\Delta\sigma_z$ from foundation loads decrease to about 10 (cohesionless soil) to 20 (cohesive soil) percent of the effective vertical overburden pressure σ_o' (item 5). Errors in settlement contributed by nonlinear heterogeneous soil below the critical depth are not significant.

APPENDIX D

ELASTIC PARAMETERS

D-1. General

The magnitudes of soil elastic distortion or immediate settlement for practical applications are evaluated from the elastic soil parameters, Young's modulus E_s, shear modulus G_s, and Poisson's ratio v_s. For most practical applications the foundation soil is heterogeneous or multilayered in which the elastic parameters can vary significantly from layer to layer.

D-2. Elastic Young's Modulus

Young's elastic modulus is commonly used for estimation of settlement from static loads. Suitable values of the elastic modulus E_s as a function of depth may be estimated from empirical correlations, results of laboratory tests on undisturbed specimens, and results of field tests.

A. DEFINITION. Materials that are truly elastic obey Hooke's law, in which each equal increments of applied uniaxial stress σ_z cause a proportionate increase in strain ϵ_z

$$\epsilon_z = \frac{1}{E} \cdot \sigma_z \qquad \text{(D-1)}$$

where E is Young's modulus of elasticity (Table D-1). Figure D-1 illustrates the stress path for the uniaxial (UT) and other test methods. An elastic material regains its initial dimensions following removal of the applied stress.

(1) Application to Soil. Hooke's law, which is applicable to homogeneous and isotropic materials, was originally developed from the observed elastic behavior of metal bars in tension. Soil is sometimes assumed to behave linearly elastic under relatively small loads. A partially elastic material obeys Hooke's law during loading, but this material will not gain its initial dimensions following removal of the applied stress. These materials are nonlinear and include most soils, especially foundation soil supporting heavy structures that apply their weight only once.

(2) Assumption of Young's Elastic Modulus. Soils tested in a conventional triaxial compression (CTCT) device under constant lateral stress will yield a tangent elastic modulus E_t equivalent to Young's

modulus. The soil modulus E_s is assumed approximately equal to Young's modulus in practical applications of the theory of elasticity for computation of settlement.

(3) Relationship with Other Elastic Parameters. Table D-2 relates the elastic modulus E with the shear modulus G, bulk modulus K and constrained modulus E_d. These parameters are defined in Table D-1.

B. EMPIRICAL CORRELATIONS. The elastic undrained modulus E_s for clay may be estimated from the undrained shear strength C_u by

$$E_s = K_c C_u \qquad \text{(D-2)}$$

where E_s = Young's soil modulus in tsf; K_c = correlation factor (Figure D-2); and C_u = undrained shear strength in tsf.

The values of K_c as a function of the overconsolidation ratio and plasticity index PI have been determined from field measurements and are therefore not affected by soil disturbance compared with measurements on undisturbed soil samples. Table D-3 illustrates some typical values for the elastic modulus.

C. LABORATORY TESTS ON COHESIVE SOIL. The elastic modulus is sensitive to soil disturbance, which may increase pore water pressure and therefore decrease the effective stress in the specimen and reduce the stiffness and strength. Fissures, which may have little influence on field settlement, may reduce the measured modulus compared with the in situ modulus if confining pressures are not applied to the soil specimen.

(1) Initial Hyperbolic Tangent Modulus. Triaxial unconsolidated undrained (Q or UU) compression tests may be performed on the best available undisturbed specimens at confining pressures equal to the total vertical overburden pressure σ_o for that specimen when in the field using the Q test procedure described in EM 1110-2-1906, Laboratory Soils Testing. An appropriate measure of E_s is the initial tangent modulus $E_{ti} = 1/a$ where a is the intercept of a plot of strain/deviator stress versus strain (Figure D-3) (item 14).

(2) Reload Modulus. A triaxial consolidated undrained (R or CU) compression test may be performed on the best available undisturbed speci-

Table D-1. Laboratory Tests for Evaluation of Elastic Parameters
(Refer to Figure D-1)

Type of Test (1)	Description (2)	Diagram (3)
UNIAXIAL STRESS (UT)	LOADING σ_1 ON A SINGLE VERTICAL AXIS. $\sigma_2 = \sigma_3 = 0$. YOUNG'S MODULUS E IS DETERMINED.	$E = \dfrac{d\sigma_1}{d\epsilon_1}$
HYDROSTATIC COMPRESSION (HCT)	LOADING OCCURS ALONG THE SPACE DIAGONAL IN EQUAL INCREMENTS $\sigma_0 = \sigma_1 = \sigma_2 = \sigma_3$ AND $\epsilon_{vol} = \epsilon_1 + \epsilon_2 + \epsilon_3$ BULK MODULUS K IS DETERMINED.	$K = \dfrac{d\sigma_0}{d\epsilon_{vol}}$
SIMPLE SHEAR (SST)	AFTER HYDROSTATIC LOADING TO $\sigma_0 = \sigma_{OCT}$ $(\sigma_1 + \sigma_2 + \sigma_3)/3$ KEPT CONSTANT, BUT TWO OF THREE STRESS AXES VARIED; i.e., $\Delta\sigma_1 = -\Delta\sigma_3$, $\Delta\sigma_2 = 0$. SHEAR MODULUS G IS DETERMINED.	$G = \dfrac{d\sigma_{12}}{d\epsilon_{12}}$
CONFINED COMPRESSION (CCT)	LOADING σ_1 WHEN $E_2 = E_3 = 0$ (CONSOLIDATION TEST) CONSTRAINED MODULUS E IS DETERMINED.	$E_d = \dfrac{d\sigma_1}{d\epsilon_1}$
CONVENTIONAL TRIAXIAL COMPRESSION (CTCT)	AFTER HYDROSTATIC LOADING TO σ_0, σ_1 INCREASED WHILE $\sigma_2 = \sigma_3$ KEPT CONSTANT AT σ_0. TANGENT MODULUS E_t IN COMPRESSION DETERMINED.	$q = \sigma_1 - \sigma_0$; $E_t = \dfrac{d(\sigma_1 - \sigma_3)}{d\epsilon_1}$
CONVENTIONAL TRIAXIAL EXTENSION (CTET)	AFTER HYDROSTATIC LOADING TO σ_0, $\sigma_2 = \sigma_3$ INCREASED WHILE σ_1 KEPT CONSTANT AT σ_0. TANGENT MODULUS E_t IN EXTENSION DETERMINED.	E_t

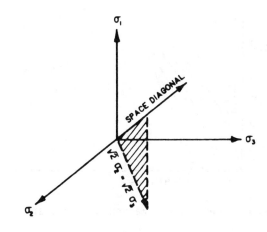

a. THE TRIAXIAL PLANE

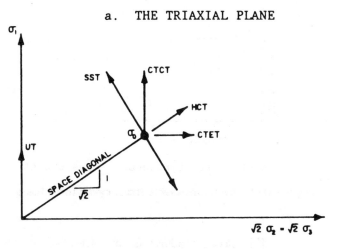

b, STRESS PATHS

Figure D-1. Examples of stress paths for different tests (refer to Table D-1 for descriptions of tests)

mens. The specimen is initially fully consolidated to an isotropic confining pressure equal to the vertical overburden pressure σ_o for that specimen in the field. The R test procedure described in EM 1110-2-1906 may be used except as follows: stress is increased to the magnitude estimated for the field loading condition.

Table D-2. Relationships Between Elastic Parameters

Parameter (1)	Relationship (2)
Shear modulus G in tsf	$\dfrac{E}{2(1+\nu)}$ $\nu = $ Poisson's ratio
Bulk modulus K in tsf	$\lambda_L + \dfrac{2G}{3}$ $\lambda_l = $ Lames's constant
Constrained modulus E_d in tsf	$\dfrac{E(1-\nu)}{(1+\nu)(1-2\nu)}$

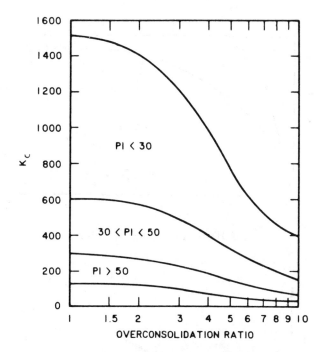

Figure D-2. Chart for estimating constant K_c to determine the elastic modulus $E_s = K_c C_u$ from the undrained shearstrength (after Figure 3-20, TM 5-818-1)

The axial stress may then be reduced to zero and the cycle repeated until the reload curve shows no further increase in slope. The tangent modulus at 1/2 of the maximum applied stress is determined for each loading cycle and plotted versus the number of cycles (Figure D-4). An appropriate measure of E_s is the reload tangent modulus that approaches the asymptotic value at large cycles.

D. FIELD TESTS. The elastic modulus may be estimated from empirical and semiempirical rela-

Table D-3. Typical Elastic Moduli

Soil (1)	$E_s q$ (tsf) (2)
a. Clay	
Very soft clay	5–50
Soft clay	50–200
Medium clay	200–500
Stiff clay, silty clay	500–1000
Sandy clay	250–2000
Clay shale	1000–2000
b. Sand	
Loose sand	100–250
Dense sand	250–1000
Dense sand and gravel	1000–2000
Silty sand	250–2000

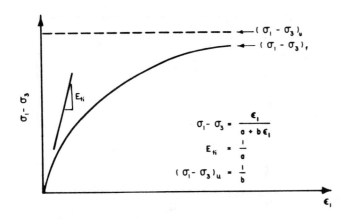

a. HYPERBOLIC RELATIONSHIP

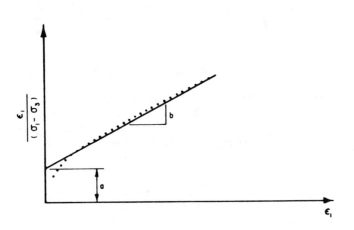

b. EVALUATION OF HYPERBOLIC PARAMETERS a, b

Figure D-3. Hyperbolic simulation of stress-strain relationships

tionships based on results of field soil tests. Refer to EM 1110-1-1804, Geotechnical Investigations, for more information on in situ tests.

(1) Plate Load Test. The plate load test performed in accordance with ASTM Standard Test Method D 1194, "Bearing Capacity of Soil for Static Loads on Spread Footings," is used to determine the relationship between settlement and plate pressure q_p (Figure D-5). The elastic modulus E_s is found from the slope of the curve $\Delta\rho/\Delta q_P$

$$E_a = \frac{(1 - \nu_s^2)}{\frac{\Delta\rho}{\Delta q_P}} \cdot B_p \cdot I_w \qquad \text{(D-3)}$$

where E_s = Young's soil modulus in psi; ν_s = Poisson's ratio, 0.4; $\Delta\rho/\Delta q_P$ = slope of settlement versus plate pressure in in./psi; B_p = diameter of plate, in in.; and I_w = influence factor, $\pi/4$ for circular plates.

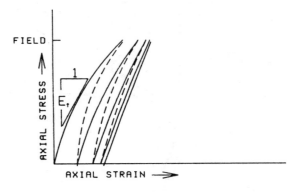

a. TANGENT MODULUS AT 1/2 MAXIMUM APPLIED STRESS

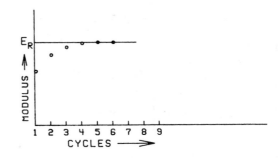

b. TANGENT RELOAD MODULUS VERSUS CYCLES

Figure D-4. Elastic modulus from cyclic load tests

This elastic modulus is representative of soil within a depth of $2B_p$ beneath the plate.

(2) Cone Penetration Test (CPT). The constrained modulus E_d has been empirically related with the cone tip bearing resistance by

$$E_d = \alpha_c \cdot q_c \qquad \text{(D-4)}$$

where E_d = constrained modulus in tsf; α_c = correlation

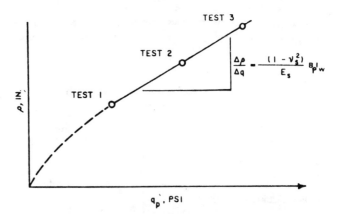

Figure D-5. Graphical solution of soil elastic modulus E_s from the plate load test; $I_w = \pi/4$ for circular rigid plate of diameter $B_p \cdot \mu_s$ = Poisson's ratio

factor depending on soil type and the cone bearing resistance (Table D-4); and q_c = cone tip bearing resistance in tsf. A typical value for sands is $\alpha_c = 3$, but it can increase substantially for overconsolidated sand. A typical value for clays is $\alpha_c = 10$ when used with the net cone resistance $q_c - \sigma_o$ where σ_o is the total overburden pressure.

The undrained shear strength C_u is related to q_c by

$$C_u = \frac{q_c - \sigma_c}{N_k} \qquad (D\text{-}5)$$

where C_u = undrained shear strength in tsf; q_c = cone tip resistance in tsf; σ_o = total overburden pressure in tsf; and N_k = cone factor. The cone factor often varies from 10 to 20 and can be greater.

(3) Standard Penetration Test (SPT). The elastic modulus in sand may be estimated directly from the blowcount by (item 60)

$$E_x = 9.4N^{0.87}\sqrt{B}\left(1 + 0.4\frac{D}{B}\right) \qquad (D\text{-}6)$$

where E_s = Young's soil modulus in tsf; N = average blowcount per ft in the stratum, number of blows of a 140-pound hammer falling 30 in. to drive a standard sampler (1.42″ ID, 2.00″ OD) 1 ft. (sampler is driven 18 in. and blows counted the last 12 in.); B = width of footing in ft; and D = depth of embedment of footing in ft.

Equation D-6 was developed from information

in the literature and original settlement observations without consideration of the energy of the hammer. An alternative method of estimating the elastic modulus for footing foundations on clean sand or sand and gravel is (after item 12)

Preloaded sand:

$$E_m = 420 + 10N_{ave} \qquad (D\text{-}7a)$$

Normally loaded sand or sand and gravel:

$$E_m = 194 + 8N_{ave} \qquad (D\text{-}7b)$$

where E_m = deformation modulus, $E_s/(1 - v_s^2)$ in tsf; and N_{ave} = average measured blowcount in depth $H = B$ below footing in blows/ft.

(4) Pressuremeter test (PMT). The preboring pressuremeter consists of a cylindrical probe of radius R_o containing an inflatable balloon lowered into a borehole to a given depth. The pressure required to inflate the balloon and probe against the side of the borehole and the volume change of the probe are recorded. The self-boring pressuremeter includes cutting blades at the head of the device with provision to permit drilling fluids to circulate and carry cuttings up to the surface. The self-boring pressuremeter should in theory lead to a less disturbed hole than the preboring pressuremeter. The pressure and volume change measurements are corrected for membrane resistance and volume losses leading to the corrected pressuremeter curve (Figure D-6). The preboring pressuremeter curve

Table D-4. Correlation Factor α_c (Data from Item 44)

Soil (1)	Resistance q_c (tsf) (2)	Water content (percent)(3)	α_c (4)
Lean clay (CL)	<7	—	3 to 8
Lean clay	7 to 20	—	2 to 5
Lean clay	>20	—	1 to 2.5
Silt (ML)	<20	—	3 to 6
Silt (ML)	>20	—	1 to 3
Plastic silt clay (CH, MH)	<20	—	2 to 6
Organic silt	<12	—	2 to 8
Organic clay peat	<7	50 to 100	1.5 to 4
Organic clay peat	<7	100 to 200	1 to 1.5
Organic clay peat	<7	>200	0.4 to 1
Sand	<50	—	2 to 4
Sand	>100	—	1.5
Sand	—	—	$1 + D_r^2$
Clayey sand	—	—	3 to 6
Silty sand	—	—	1 to 2
Chalk	20 to 30	—	2 to 4

Note: D_r = relative density (fraction).

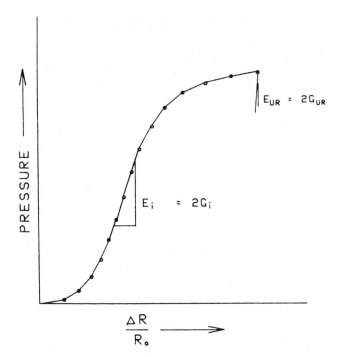

Figure D-6. Example of corrected preboring pressuremeter curve

indicates a pressuremeter modulus E_i that initially increases with increasing radial dimensional change, $\Delta R / R_o$, as shown in Figure D-6. The self-boring pressuremeter curve is characteristic of an initially high pressuremeter modulus E_i that decreases with increasing volume change without the initial increasing modulus shown in the figure. The pressuremeter modulus is a measure of twice the shear modulus. If the soil is perfectly elastic in unloading, characteristic of a sufficiently small unload-reload cycle, the gradient will be $2G_{UR}$ (item 23). The unload-reload modulus should be determined on the plastic portion of the pressuremeter curve. The pressuremeter modulus may be evaluated from the gradient of the unload-reload cycle by (ASTM 4719)

$$E_p = \frac{(1 + v_s) \cdot \Delta P \cdot (R_{po} + \Delta R_{pm})}{\Delta R_p} \quad \text{(D-8)}$$

where v_s = soil Poisson's ratio, 0.33; ΔP = change in pressure measured by the pressuremeter in tsf; R_{po} = radius of probe in in.; ΔR_{pm} = change in radius from R_{po} at midpoint of straight portion of the pressuremeter curve, in in.; ΔR_p = change in radius between selected straight portions of the pressuremeter curve in in.

E. EQUIVALENT ELASTIC MODULUS. The following two methods are recommended for calculating an equivalent elastic modulus of cohesive soil for estimating settlement of mats and footings.

(1) Kay and Cavagnaro Approximation. The equivalent elastic modulus E_s^* may be calculated by (item 31)

$$E_s^* = \frac{2 \cdot q \cdot R \cdot (1 - v_s^2)}{\rho_c} \quad \text{(D-9)}$$

where E_s^* = equivalent elastic modulus in tsf; q = bearing pressure in tsf; R = equivalent mat radius, $\sqrt{LB/\pi}$, $L < 2B$ in ft; L = length of mat in ft; B = width of mat in ft; and ρ_c = center settlement from the Kay and Cavagnaro method (Figure 3-10) in ft.

(2) Semiempirical Method. The equivalent elastic modulus of a soil with elastic modulus increasing linearly with depth may be estimated by

$$E_s^* = \frac{2 \cdot k \cdot R \cdot (1 - v_s^2)}{0.7 + (2.3 - 4 \cdot v_s) \cdot \log n} \quad \text{(D-10)}$$

where k = constant relating soil elastic modulus E_s with depth z; $E_s = E_o + kz$ in tons/ft³; D = depth of foundation below ground surface in ft; $n = kR/(E_o + kD_o)$; and E_o = elastic soil modulus at the ground surface in tsf. Equation D-10 was developed from results of a parametric study using Equation D-9 (item 29).

(3) Gibson Model. The equivalent modulus of a soil with elastic modulus increasing linearly with depth and $E_o = 0$ is (item 19)

$$E_s^* = \frac{Bk}{2} \quad \text{(D-11)}$$

where B is the minimum width of the foundation in ft.

D-3. Shear Modulus

The shear modulus G may be used for analysis of settlement from dynamic loads.

A. DEFINITION. Shear stresses applied to an elastic soil will cause a shear distortion illustrated by the simple shear test (SST) (Table D-1).

B. EVALUATION BY DYNAMIC TESTS. The shear modulus may be evaluated from dynamic tests after methodology of chapter 17, TM 5-818-1, Procedures for Foundation Design of Buildings and Other Structures (Except Hydraulic Structures).

C. RELATIONSHIPS WITH OTHER PARAMETERS. Table D-2 illustrates the relationship of the shear modulus with Young's elastic E and bulk modulus K.

D-4. Poisson's Ratio

A standard procedure for evaluation of Poisson's ratio for soil does not exist. Poisson's ratio v_s for soil usually varies from 0.25 to 0.49 with saturated soils approaching 0.49. Poisson's ratio for unsaturated soils usually vary from 0.25 to 0.40. A reasonable overall value for v_s is 0.40. Normal variations in elastic modulus of foundation soils at a site are more significant in settlement calculations than errors in Poisson's ratio.

APPENDIX E

LABORATORY CONSOLIDOMETER TESTS

E-1. General

The following laboratory tests, which may be used to evaluate consolidation parameters of compressible soils, are described. The step load (SL) and rapid load (RL) tests are normally recommended. The constant rate of deformation (CRD) and controlled gradient (CG) tests were developed to reduce the time required to complete the test relative to the SL test.

E-2. Step Load Test (SL)

This test, described in chapter VIII, EM 1110-2-1906, Laboratory Soils Testing, and ASTM D 2435, One-Dimensional Consolidation Properties of Soils, may be performed with fixed or floating ring consolidometers (Figure E-1). A uniform vertical loading pressure is applied on the loading plate in increments to a thin specimen. The specimen should not be less than 2.00 in. in diameter by 0.5 in. in height. The decrease in height of the specimen following axial drainage of water from the specimen for each pressure increment is monitored with time. The duration of each pressure increment is usually 24 hours. Lateral strain is prevented by the specimen ring.

A. SPECIMEN PREPARATION. Undisturbed specimens shall be trimmed in a humid room to prevent evaporation of soil moisture. A glass plate may be placed on top of the specimen and the specimen gently forced into the specimen ring during the trimming operation. A cutting tool is used to trim the specimens to accurate dimensions. The top and bottom surfaces of the specimen are trimmed flush with the specimen ring. Compacted specimens may be compacted directly into a mold that includes the specimen ring.

B. TEST PROCEDURE. The specimen and specimen ring are weighed and then assembled in the loading device (Figure E-1). An initial dial indicator reading is taken. The seating pressure from the top porous stone and loading plate should not exceed 0.01 tsf. The specimen is inundated after a pressure of 0.25 tsf is applied to the specimen; additional load increments should be applied if the specimen swells until all

swelling ceases. Loading increments are applied to the specimen in increments of 0.25, 0.5, 1.0, 2.0, 4.0, 8.0, and 16.0 tsf. Each load increment should remain a minimum of 24 hours or until primary consolidation is completed. The data may be plotted as illustrated in Figure 3-17.

C. POSSIBLE ERRORS. Sources of error in consolidation results include sample disturbance, specimen not completely filling the ring, too low permeability of porous stones, friction between the specimen and specimen ring, and unsatisfactory height of the specimen.

(1) Friction. Side friction may be reduced by using larger-diameter specimens, thinner specimens, or lining the consolidation ring with teflon.

(2) Specimen Height. The specimen thickness determines how clearly the break in the consolidation-time curve represents completion of primary consolidation. Specimens that are too thin may cause the time to 100 percent consolidation to be too rapid. The break in the curve indicating end of primary consolidation may be obscured by secondary compression if the specimen is too thick.

D. TIME REQUIREMENTS. A properly conducted test may take several weeks or months and may be especially time-consuming for soft or impervious soil.

E-3. Rapid Load Test (RL)

The RL test is similar to the SL test except much larger pressure increment ratios may be used and the duration of each pressure increment is restricted. The time duration is usually limited to allow only 90 percent of full consolidation as evaluated by Taylor's square root of time method (Table 3-11). Refer to item 47 for further details.

A. TIME REQUIREMENTS. This test may be performed in a single day.

B. ACCURACY. Accuracy is similar to the SL procedure.

C. PRESSURE INCREMENTS. Large pressure increments exceeding those of the SL test for

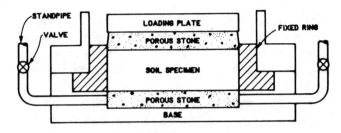

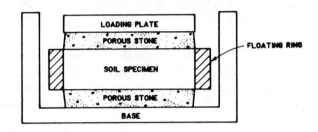

Figure E-1. Schematic diagrams of fixed-ring and floating-ring consolidometers

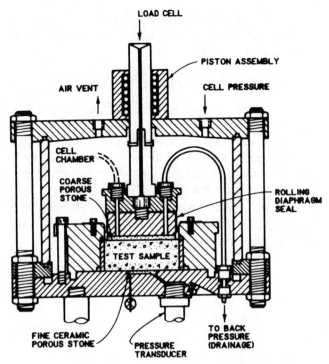

Figure E-2. General purpose consolidometer. Reprinted by permission of the American Society of Civil Engineers from the _Journal of the Soil Mechanics and Foundations Division_, Vol 97, 1971, "Consolidation at Constant Rate of Strain", by A. E. Z. Wizza, J. T. Christian, and E. H. Davis, p 1394

applied pressures exceeding the maximum past pressure reduces the amount of secondary compression contained in the void ratio-logarithm time curve (Figure 3-17).

E-4. Constant Rate of Deformation Test (CRD)

A thin cylindrical soil specimen similar to that of the SL test is saturated at constant volume under a backpressure and loaded vertically without lateral strain at a constant rate of vertical strain. Drainage is permitted only from the upper surface of the specimen. A general-purpose consolidometer capable of this test procedure is shown in Figure E-2.

A. EVALUATION OF MAXIMUM PAST PRESSURE. The maximum past pressure determined by this procedure is dependent on the rate of strain and increases with increase in strain rate. The strain rate should be consistent with expected field rates. Typical field strain rates are about 10^{-7} per minute.

B. EVALUATION OF COEFFICIENT OF CONSOLIDATION. c_v should be evaluated by (item 67)

$$c_v = \frac{h^2 \cdot \log_{10} \dfrac{\sigma_2}{\sigma_1}}{2 \cdot \Delta t \cdot \log_{10} \left[1 - \dfrac{u_h}{\sigma_a} \right]} \qquad \text{(E-1)}$$

where h = average thickness of drainage path (specimen thickness) in in.; σ_1 = total vertical stress at time t_1 in psi; σ_2 = total vertical stress at time t_2 in psi; $\Delta t = t_2 - t_1$ in min; u_h = average excess pore water pressure at the bottom of the specimen over the time interval t_1 and t_2 in psi; and $\sigma_a = (\sigma_1 + \sigma_2)/2$, average total vertical stress over time interval t_1 and t_2 in psi. The coefficient of consolidation c_v determined by this method appears comparable with that of the SL test.

C. EVALUATION OF VOID RATIO–LOGARITHM PRESSURE RELATIONSHIP. The void ratio–logarithm pressure relationship may be obtained by determining the void ratio and effective stress at any time during the test.

(1) The change in void ratio Δe over a pressure increment is $(1 + e_1)\epsilon$ where e_1 is the void ratio at the beginning of the pressure increment and ϵ is the strain over the pressure increment.

(2) The average effective stress over a pressure increment is $\sigma_a - u_h$.

(3) The excess pore water pressure u_h is measured at the bottom of the specimen (Figure E-2).

D. ASSUMPTIONS. This method assumes that the coefficient of consolidation and compression index are constant for the soil.

E-5. Controlled Gradient Test (CG)

This test is similar to the CRD test except that the applied vertical pressure is adjusted so that the pore water pressure at the bottom of the specimen remains constant throughout the test (item 38). This restriction requires a feedback mechanism that significantly complicates the laboratory equipment.

A. EVALUATION OF COEFFICIENT OF CONSOLIDATION. The coefficient of consolidation may be estimated by

$$c_v = \frac{\Delta \sigma \cdot h_a^2}{\Delta t \cdot 2u_h} \qquad (E\text{-}2)$$

where $\Delta\sigma$ = change in total pressure between time increment t_2 and t_1 in psi; $\Delta t = t_2 - t_1$, min; h_a = average height of specimen between time t_2 and t_1 in in.; and u_h = excess pore water pressure at bottom of the specimen in psi.

B. EVALUATION OF VOID RATIO–LOGARITHM PRESSURE RELATIONSHIP. The void ratio–logarithm pressure relationship is evaluated similar to the above procedures. The excess pore water pressure at the bottom of the specimen should be kept as small as possible to maintain a nearly uniform void ratio within the specimen.

C. ASSUMPTIONS. This method assumes that the coefficient of consolidation and coefficient of volume change are constant. The coefficient of volume change is the change in strain divided by the change in total vertical stress.

APPENDIX F

COMPUTER PROGRAM VDISPL

F-1. Background

This program, Vertical DISPLacements, was developed to assist in the calculation of vertical displacements beneath shallow foundations for various types of multilayered soils in support of this manual. The two types of foundations considered are rectangular footings or mats and long strip footings. These foundations are assumed flexible. Models available are immediate settlement of granular soil from cone penetration data (item 55), immediate settlement of granular soil from both cone penetration and dilatometer data (item 37), immediate settlement of an elastic soil using the Boussinesq pressure distribution and Young's elastic soil modulus, consolidation or swell of a cohesive soil (ASTM D 4546), and settlement of a collapsible soil (item 22). Soil expansion is indicated by positive values and settlement by negative values.

F-2. Organization

The program consists of a main routine and several subroutines for calculation of vertical displacements by each of the models. The main routine feeds in descriptive data for the problem, divides the soil profile into increments from 1 to NNP where NNP is the nodal point at the bottom of the profile, assigns a layer number to each increment or soil element, and calculates the effective overburden pressure prior to placement of the foundation. The number of soil elements NEL in a problem is one less than the number of nodal points NNP. The PARAMETER statement in the program provides a maximum NL=30 soil layers and a maximum NQ=101 soil nodal points. Subroutine SLAB calculates the change in effective vertical pressures in the soil following placement of the foundation using the Boussinesq pressure distribution. The remaining subroutines described later calculate vertical displacements for the various models.

A. DATA ORGANIZATION. The input data are placed in a file, "VDIN.DAT." These data are illustrated in Table F-1 with descriptions provided in Table F-2. The input data are also printed in the output file, "VDOU.DAT" and illustrated in Table F-3. Calculations by the program are printed in VDOU.DAT and illustrated in Table F-3.

B. SUBROUTINE MECH. This subroutine (NOPT=0) calculates heave or consolidation from results of one-dimensional consolidometer swell tests performed on cohesive soil for each layer in a soil profile.

(1) Soil displacements are calculated after ASTM standard D 4546 (method C) relative to the equilibrium moisture profiles of saturated (method 1), hydrostatic with shallow water table (method 2), and hydrostatic without shallow water table (method 3) (Figure 5-1).

(2) Input data include swell pressure, swell and compression indices, and the maximum past pressure of each soil layer. Incremental and total settlement in soil adjacent to the foundation and below the foundation may be calculated.

C. SUBROUTINE LEON. This subroutine (NOPT=1) calculates immediate settlement of shallow foundations on granular soils using both cone penetration and dilatometer in situ test data. This method considers effects of prestress after the model of Leonards and Frost (item 37)

$$\rho_i = C_i \cdot \Delta p \cdot \Delta z \cdot \sum_{i=1}^{i=NEL} I_{iz} \left[\frac{R_{izoc}}{E_{izoc}} + \frac{R_{iznc}}{E_{iznc}} \right] \quad \text{(F-1)}$$

where ρ_i = immediate settlement in ft; c_1 = correction to account for strain relief from embedment, $1 - 0.5\sigma_o'/\Delta p \geq 0.5$; σ_o' = effective vertical overburden pressure at bottom of footing in tsf; Δp = net applied footing pressure, $q - \sigma_o'$ in tsf; q = bearing pressure on footing in tsf; Δz = depth increment in ft; I_{iz} = influence factor of soil layer i from Figure 3-4; R_{izoc} = ratio of stress increment corresponding to the overconsolidated part in soil layer i to the total stress increment in that layer; R_{iznc} = ratio of stress increment corresponding to the normally consolidated part in soil layer i to the total stress increment in that layer; E_{izoc} = Young's soil modulus of overconsolidated soil layer i at depth z in tsf; and E_{iznc} = Young's soil modulus of normally consolidated soil layer i at depth z in tsf.

(1) The ratios R_{iz} may be found as follows:

$$R_{izoc} = \frac{\sigma_p' - \sigma_v'}{\sigma_f' - \sigma_v'} \quad \text{(F-2a)}$$

$$R_{iznc} = \frac{\sigma_f' - \sigma_p'}{\sigma_p' - \sigma_v'} \quad \text{(F-2b)}$$

Table F-1. Input Data

Line (1)	Input parameters (2)	Format statement (3)
1	TITLE	20A4
2	MPROB NOPT NBPRES NNP NBX NMAT DX	6I5,F10.2
3	N IE(N,M) (Line 3 repeated for each new material, M=1,NMAT; last line is NEL NMAT)	2I5
4	M G(M) WC(M) EO(M) (Line 4 repeated for each new material, M=1,NMAT)	I5,3F10.3
5	DGWT IOPTION NOUT	F10.2,2I5
6	Q BLEN BWID MRECT	3F10.2,I5
7	If NOPT=0 M SP(M) CS(M) CC(M) PM(M)	I5,4F10.4
	If NOPT=1 M PO(M) P1(M) QC(M)	I5,3F10.2
	If NOPT=2 M QC(M)	I5,F10.2
	If NOPT=3 M PRES(M,J),J=1,5	I5,5F10.2
	If NOPT=4 M ES(M)	I5,F10.2
	(Line 7 repeated for each new material, M=1,NMAT)	
8	If NOPT=0 XA XF	2F10.2
	If NOPT=2 or 4 TIME	F10.2
	If NOPT=3 M STRA(M,J),J=1,5 (This line repeated for each new material, M=1,NMAT)	I5,5F10.2

Table F-2. Description of Input Data

Line (1)	Parameter (2)	Description (3)
1	TITLE	Name of problem
2	NPROB	Number of problems; If NPROB >1, then the program repeats calculations with new input data beginning with Line 5
	NOPT	Option for model: =0 for consolidation/swell =1 for Leonards and Frost =2 for Schmertmann =3 for Collapsible soil =4 for elastic soil settlement
	NBPRES	Option for foundation: =1 for rectangular slab =2 for long strip footing
	NNP	Total number of nodal points; vertical displacements calculated to depth DX*(NNP-1)
	NBX	Number of nodal point at bottom of foundation
	NMAT	Total number of different soil layers
	DX	Increment of depth, ft
3	N	Element number
	IE(N,M)	Number of soil layer M associated with element N
4	M	Number of soil layer
	G(M)	Specific gravity of soil layer M
	WC(M)	Water content of soil layer M, percent
	EO(M)	Initial void ratio of soil layer M
5	DGWT	Depth to hydrostatic water table, ft; If NOPT=0 and IOPTION =1, then set DGWT=ZA+Uwa/0.03125 where Uwa= suction (or positive value of negative pore water pressure),tsf at depth ZA (method 3, Figure 5-1); If Uwa=0, then DGWT=ZA (method 2, Figure 5-1); set NNP= 1+ ZA/DX to prevent displacement calculations below depth ZA
	IOPTION	Equilibrium moisture profile: =0 for saturation above the water table (method 1, Figure 5-1); if NOPT=0, then =1 for hydrostatic above water table
	NOUT	Amount of output data, =0 only total displacements =1 for increments with totals
6	Q	Applied uniform pressure on foundation, tsf
	BLEN	Length of foundation or 0.0 if NBPRES=2, ft
	BWID	Width of foundation or long continuous footing, ft
	MRECT	Location of calculation: =0 for center and =1 for corner of rectangle or edge of long footing

Table F-2. Concluded

Line (1)	Parameter (2)	Description (3)
7		If NOPT=0
	M	Number of soil layer
	SP(M)	Swell pressure of soil layer M, tsf
	CS(M)	Swell index of soil layer M
	CC(M)	Compression index of soil layer M
	PM(M)	Maximum past pressure of soil layer M, tsf
		If NOPT=1
	M	Number of soil layer
	PO(M)	Dilatometer A-pressure, tsf
	P1(M)	Dilatometer B-pressure, tsf
	QC(M)	Cone penetration resistance, tsf
		If NOPT=2
	M	Number of soil layer
	QC(M)	Cone penetration resistance, tsf
		If NOPT=3
	M	Number of soil layer
	PRES(M,5)	Applied pressure at 5 points of collapsible soil test, tsf; must be greater than zero
		If NOPT=4
	M	Number of soil layer
	ES(M)	Elastic modulus, tsf
8		If NOPT=0
	ZA	Depth of active zone of heave, ft
	XF	Depth from ground surface to the depth that heave begins, ft
		If NOPT=2 or 4
	TIME	Time in years after construction for Schmertmann model
		If NOPT=3
	M	Number of soil layer
	STRA(M,5)	Strain at 5 points of collapsible soil test, percent

Table F-3. Output Data

Line (1)	Output parameters (2)	FORTRAN statement (3)
1	TITLE	20A4
2	NUMBER OF PROBLEMS= NUMBER OF NODAL POINTS=	I5 I5
3	NUMBER OF NODAL POINT AT BOTTOM OF FOUNDATION=	I11
4	NUMBER OF DIFFERENT SOIL LAYERS= INCREMENT DEPTH=	I5 F10.2
5	If NOPT=0 CONSOLIDATION SWELL MODEL If NOPT=1 LEONARDS AND FROST MODEL If NOPT=2 SCHMERTMANN MODEL If NOPT=3 COLLAPSIBLE SOIL If NOPT=4 ELASTIC SOIL	
6	If NBPRES=1 RECTANGULAR SLAB FOUNDATION If NBPRES=2 LONG CONTINUOUS STRIP FOUNDATION	
7	DEPTH OF FOUNDATION= FEET	F10.2
8	TOTAL DEPTH OF SOIL PROFILE= FEET	F10.2
9	ELEMENT NUMBER OF SOIL	I5,8X,I5
10	MATERIAL SPECIFIC GRAVITY WATER CONTENT, % VOID RATIO	I5,3F10.3
11	DEPTH TO WATER TABLE= FEET	F10.2
12	If NOUT=0 TOTAL DISPLACEMENTS ONLY If NOUT=1 DISPLACEMENTS AT EACH DEPTH OUTPUT	

Table F-3. Continued

Line (1)	Output parameters (2)	FORTRAN statement (3)
13	If IOPTION=0 or NOPT=0 EQUILIBRIUM SATURATED ABOVE WATER TABLE If IOPTION=1 and NOPT=0 EQUILIBRIUM HYDROSTRATIC PROFILE ABOVE WATER TABLE	
14	APPLIED PRESSURE ON FOUNDATION= TSF	F10.2
15	LENGTH= FEET WIDTH= FEET	F10.2,F10.2
16	If MRECT=0 CENTER OF FOUNDATION If MRECT=1 CORNER OF SLAB OR EDGE OF LONG STRIP FOOTING	

If NOPT=0

Line	Output parameters	FORTRAN statement
17	MATERIAL SWELL PRESSURE, SWELL COMPRESSION MAXIMUM PAST TSF INDEX INDEX PRESSURE,TSF	I5,4F15.3
18	ACTIVE ZONE DEPTH (FT)= DEPTH ACTIVE ZONE BEGINS (FT)=	F10.2 F10.2
19A	If NOUT=1 HEAVE DISTRIBUTION ABOVE FOUNDATION DEPTH ELEMENT DEPTH,FT DELTA HEAVE,FT EXCESS PORE PRESSURE,TSF	I5,F13.2,2F18.5
19B	HEAVE DISTRIBUTION BELOW FOUNDATION DEPTH ELEMENT DEPTH,FT DELTA HEAVE,FT EXCESS PORE PRESSURE,TSF	
20	SOIL HEAVE NEXT TO FOUNDATION EXCLUDING HEAVE IN SUBSOIL BENEATH FOUNDATION= FEET SUBSOIL MOVEMENT= FEET TOTAL HEAVE FEET	F8.5 F8.5 F8.5

If NOPT=1

Line	Output parameters	FORTRAN statement
17	MATERIAL A PRESSURE, TSF B PRESSURE, TSF CONE RESISTANCE, TSF	I5,3F18.2
18	If NOUT=1 ELEMENT DEPTH, SETTLEMENT, KO QC/SIGV PHI,DEGREES FT FT	I5,F10.2,F13.5,3F10.2

Table F-3. Continued

Line (1)	Output parameters (2)	FORTRAN statement (3)
19	SETTLEMENT BENEATH FOUNDATION= FEET	F10.5

If NOPT=2

17	MATERIAL CONE RESISTANCE, TSF	I5,F18.2
18	TIME AFTER CONSTRUCTION IN YEARS=	F10.2
19	If NOUT=1 ELEMENT DEPTH, FT SETTLEMENT, FT	I5,F13.2,F18.5
20	SETTLEMENT BENEATH FOUNDATION= FEET	F10.5

If NOPT=3

17	APPLIED PRESSURE AT 5 POINTS IN UNITS OF TSF MATERIAL A BB B C D	I5,5F10.2
18	STRAIN AT 5 POINTS IN PERCENT MATERIAL A BB B C D	I5,5F10.2
19	If NOUT=1 COLLAPSE DISTRIBUTION ABOVE FOUNDATION DEPTH ELEMENT DEPTH,FT DELTA,FT	I5,F13.2,F18.2
	COLLAPSE DISTRIBUTION BELOW FOUNDATION DEPTH ELEMENT DEPTH,FT DELTA,FT	I5,F13.2,F18.2
20	SOIL COLLAPSE NEXT TO FOUNDATION EXCLUDING COLLAPSE IN SUBSOIL BENEATH FOUNDATION FEET SUBSOIL COLLAPSE= FEET TOTAL COLLAPSE= FEET	F10.5 F10.5 F10.5

If NOPT=4

17	MATERIAL ELASTIC MODULUS, TSF	I5,F18.2
18	TIME AFTER CONSTRUCTION IN YEARS=	
19	If NOUT=1 SETTLEMENT DISTRIBUTION BELOW FOUNDATION ELEMENT DEPTH,FT DELTA,FT	I5,F13.2,F18.5
20	SUBSOIL SETTLEMENT BENEATH FOUNDATION FEET	F10.5

where σ_p' = preconsolidation or maximum past pressure in tsf; σ_v' = initial vertical effective stress in tsf; and σ_f' = final effective stress at the center of the layer in tsf. σ_p' is determined from σ_v' and the overconsolidation ratio (OCR).

(2) The OCR is estimated from (item 57)

$$OCR = \left[\frac{K_{oc}}{1 - \sin \phi_{ax}} \right]^{\frac{1}{0.8 \cdot \sin\phi_{ax}}} \qquad \text{(F-3)}$$

where ϕ_{ax} is the axial friction angle in degrees. Equation F-3 was confirmed (item 37) for OCR between 2 and 20.

(3) ϕ_{ax} is estimated from the plane strain angle ϕ_{ps} by (item 57)

$$\phi_{ax} = \phi_{px} - \left[\frac{\phi_{ps} - 32}{3} \right] \qquad \text{(F-4)}$$

(4) The plane strain angle ϕ_{ps} is estimated from the coefficient of earth pressure and q_c/σ_v' using correlations suggested by (item 41). q_c is the cone penetration resistance in tsf. These correlations have been programmed in subroutine SPLINE. Subroutine BICUBIC is used to evaluate ϕ_{ps}. The coefficient of earth pressure is estimated using

$$K_{oc} = 0.376 + 0.095K_P - 0.0017 \, (q_c/\sigma_v') \quad \text{(F-5)}$$

where K_{oc} = coefficient of earth pressure for overconsolidated soil; K_D = horizontal stress index, $(P_o - u_w)/\sigma_v'$; P_o = dilatometer lift-off pressure in tsf; and u_w = in situ pore water pressure in tsf.

Settlements calculations are sensitive to the value of $K_{D\prime}$, the horizontal stress index. If K_{oc} calculated by Equation F-5 exceeds 1.8, then the OCR calculated by Equation E-3 may exceed 20. The dilatometer lift-off pressure P_o should be a reliable value. Program VDISPL does not limit the value of K_{oc}. The constants in Equation F-5 may also differ from those for local soils.

(5) The elastic soil moduli are estimated by

$$E_{izoc} = 3.5E_D \qquad \text{(F-6a)}$$

$$E_{iznc} = 0.7E_D \qquad \text{(F-6b)}$$

where E_D = dilatometer modulus in tsf. E_D is given by

$$E_D = 34.7(P_1 - P_o) \qquad \text{(F-7)}$$

where P_1 is the dilatometer B-pressure, the pressure required to expand the membrane 1.1 mm at the test depths. If $R_{izoc} = 0$, then $E_{iznc} = 0.9E_D$. Input data include

the dilatometer A and B pressures and the cone resistance for each soil layer. Incremental and total settlement beneath the foundation may be calculated.

D. SUBROUTINE SCHMERT. This subroutine (NOPT=2) calculates immediate settlement of shallow foundations on granular soils from cone penetration test data by the Schmertmann model (item 55)

$$\rho_i = C_i \cdot C_t \cdot \Delta p \cdot \Delta z \cdot \sum_{i=1}^{NEL} \frac{I_{iz}}{E_{si}} \qquad \text{(F-8)}$$

where the input parameters are the same as above except E_{si} is the elastic soil modulus of layer i. $E_{si} = 2.5q_c$ for rectangular footings or mats and $E_{si} = 3.5q_c$ for long strip footings. C_t is a correction for time-dependent increase in settlement

$$C_t = 1 + 0.2 \cdot \log_{10} (t/0.1) \qquad \text{(F-9)}$$

(1) Input data include cone penetration resistance q_c for each soil layer and the time in years following construction when settlement is to be calculated. The incremental and total settlement beneath the foundation is calculated for the provided time.

(2) This subroutine includes an option (NOPT=4) to input E_{si} directly for each soil layer. This option may be useful if the user does not have cone penetration resistance data but can determine reliable values of E_{si} by some other tests.

E. SUBROUTINE COLL. This subroutine (NOPT=3) calculates the collapse settlement of susceptible soils after the model of Houston et al. (item 22). The model uses the results of a one-dimensional consolidometer test performed on collapsible soil for each soil layer (Figure 5-5). Input data include five points each of the applied pressure and strain distributions. The collapse settlement is calculated from the difference in strains between the unwetted line, points A-BB-B, and the wetted line, points A-C-D. Incremental and total settlement in soil adjacent to and above the foundation and below the foundation may be calculated. The settlement caused by foundation loads prior to collapse may be estimated by the Leonards and Frost (NOPT=1, item 37) and Schmertmann (NOPT=2) models (item 57).

F-3. Example Problems

A. HEAVE OF EXPANSIVE SOIL. A footing 3 ft by 3 ft square, $B = W = 3$ ft, is to be constructed 3 ft below ground surface, $D = 3$ ft, on two cohesive expansive soil layers. The amount of heave is to be calculated if the soil is left untreated. The bottom of the soil profile is 8 ft below ground surface, which is also the depth to the groundwater level. The depth increment DX is taken as 0.5 ft. The total number of

Table F-4. Expansive Soil Heave

a. Input Data, File VDIN.DAT

```
    FOOTING IN EXPANSIVE SOIL
   1    0    1   17    7    2       0.50
   1    1
  12    2
  16    2
   1    2.700    20.000       1.540
   2    2.65     19.300       0.900
      8.00    0    1
      1.00       3.00       3.00    0
   1    2.0000    0.1500    0.2500
   2    3.0000    0.1000    0.2000
      8.00         0
```

b. Output Data, File VDOU.DAT

```
    FOOTING IN EXPANSIVE SOIL

NUMBER OF PROBLEMS=    1      NUMBER OF NODAL POINTS=    17
NUMBER OF NODAL POINT AT BOTTOM OF FOUNDATION=          7
NUMBER OF DIFFERENT SOIL LAYERS=   2     INCREMENT DEPTH=      0.50  FT

        CONSOLIDATION SWELL MODEL

        RECTANGULAR SLAB FOUNDATION

DEPTH OF FOUNDATION =                3.00   FEET
TOTAL DEPTH OF THE SOIL PROFILE =    8.00   FEET
ELEMENT      NUMBER OF SOIL

      1             1
      2             1
      3             1
      4             1
      5             1
      6             1
      7             1
      8             1
      9             1
     10             1
     11             1
     12             2
     13             2
     14             2
     15             2
     16             2
```

Table F-4. Continued

b. Output Data, File VDOU.DAT

MATERIAL	SPECIFIC GRAVITY	WATER CONTENT, %	VOID RATIO
1	2.700	20.000	1.540
2	2.650	19.300	0.900

DEPTH TO WATER TABLE =   8.00   FEET

DISPLACEMENTS AT EACH DEPTH OUTPUT

EQUILIBRIUM SATURATED PROFILE ABOVE WATER TABLE

APPLIED PRESSURE ON FOUNDATION=   1.00 TSF
LENGTH =   3.00   FEET   WIDTH =   3.00   FEET

            CENTER OF FOUNDATION

MATERIAL	SWELL PRESSURE, TSF	SWELL INDEX	COMPRESSION INDEX	MAXIMUM PAST PRESSURE,TSF

SWELL PRESSURE   2.00 WAS SET GREATER THAN MAXIMUM PAST PRESSURE   0.00

WHICH IS NOT POSSIBLE; SWELL PRESSURE SET EQUAL TO MAXIMUM PAST PRESSURE

| 1 | 2.000 | 0.150 | 0.250 | 2.000 |

SWELL PRESSURE   3.00 WAS SET GREATER THAN MAXIMUM PAST PRESSURE   0.00

WHICH IS NOT POSSIBLE; SWELL PRESSURE SET EQUAL TO MAXIMUM PAST PRESSURE

| 2 | 3.000 | 0.100 | 0.200 | 3.000 |

        ACTIVE ZONE DEPTH (FT) =   8.00
DEPTH ACTIVE ZONE BEGINS (FT) =   0.00

HEAVE DISTRIBUTION ABOVE FOUNDATION DEPTH

ELEMENT	DEPTH,FT	DELTA HEAVE,FT	EXCESS PORE PRESSURE,TSF
1	0.25	0.13598	1.99003
2	0.75	0.10780	1.97010
3	1.25	0.09470	1.95017
4	1.75	0.08607	1.93024
5	2.25	0.07962	1.91031
6	2.75	0.03312	1.45017

Table F-4. Concluded

HEAVE DISTRIBUTION BELOW FOUNDATION

ELEMENT	DEPTH,FT	DELTA HEAVE,FT	EXCESS PORE PRESSURE,TSF
7	3.25	0.01780	1.00071
8	3.75	0.01886	1.04123
9	4.25	0.02171	1.14230
10	4.75	0.02552	1.26050
11	5.25	0.02926	1.36090
12	5.75	0.03802	2.43160
13	6.25	0.03987	2.47576
14	6.75	0.04105	2.50203
15	7.25	0.04167	2.51531
16	7.75	0.04185	2.51923

```
SOIL HEAVE NEXT TO FOUNDATION EXCLUDING HEAVE
IN SUBSOIL BENEATH FOUNDATION = 0.26864   FEET

          SUBSOIL MOVEMENT = 0.15780   FEET
               TOTAL HEAVE = 0.42645   FEET
```

Table F-5. Settlement of Granular Soil, Leonard and Frost Model

a. Input Data, File VDIN.DAT

```
       FOOTING IN GRANULAR SOIL — LEONARD AND FROST
    1    1    1    17    7    2      0.50
    1    1
   12    2
   16    2
    1    2.700    20.000    1.540
    2    2.65     19.300    0.900
      8.00    0    1
      1.00      3.00      3.00    0
    1    3.00      15.00      70.00
    2    5.00      20.00     100.00
```

b. Output Data, File VDOU.DAT

```
       FOOTING IN GRANULAR SOIL — LEONARD AND FROST

NUMBER OF PROBLEMS=    1      NUMBER OF NODAL POINTS=    17
NUMBER OF NODAL POINT AT BOTTOM OF FOUNDATION=         7
NUMBER OF DIFFERENT SOIL LAYERS=    2      INCREMENT DEPTH=      0.50   FT

          LEONARDS AND FROST MODEL

          RECTANGULAR SLAB FOUNDATION

DEPTH OF FOUNDATION =              3.00   FEET
TOTAL DEPTH OF THE SOIL PROFILE =  8.00   FEET
ELEMENT     NUMBER OF SOIL

    1            1
    2            1
```

Table F-5. Continued

b. Output Data, File VDOU.DAT

ELEMENT	NUMBER OF SOIL
3	1
4	1
5	1
6	1
7	1
8	1
9	1
10	1
11	1
12	2
13	2
14	2
15	2
16	2

MATERIAL	SPECIFIC GRAVITY	WATER CONTENT, %	VOID RATIO
1	2.700	20.000	1.540
2	2.650	19.300	0.900

DEPTH TO WATER TABLE =   8.00   FEET

DISPLACEMENTS AT EACH DEPTH OUTPUT

EQUILIBRIUM SATURATED PROFILE ABOVE WATER TABLE

APPLIED PRESSURE ON FOUNDATION=   1.00 TSF
LENGTH =   3.00   FEET   WIDTH =   3.00   FEET

      CENTER OF FOUNDATION

MATERIAL	A PRESSURE, TSF	B PRESSURE, TSF	CONE RESISTANCE, TSF
1	3.00	15.00	70.00
2	5.00	20.00	100.00

QNET=   0.88041

ELEMENT	DEPTH, FT	SETTLEMENT, FT	KO	QC/SIGV	PHI, DEGREES
7	3.25	−0.00017	1.66	540.32	44.52
8	3.75	−0.00033	1.49	468.28	44.10
9	4.25	−0.00053	1.36	413.19	43.67
10	4.75	−0.00068	1.25	369.69	43.26
11	5.25	−0.00071	1.17	334.49	42.88
12	5.75	−0.00152	1.69	430.59	43.68
13	6.25	−0.00150	1.56	387.24	43.32
14	6.75	−0.00142	1.45	351.82	42.98
15	7.25	−0.00127	1.36	322.33	42.67
16	7.75	−0.00104	1.28	297.41	42.39

SETTLEMENT BENEATH FOUNDATION= −0.00918   FEET

nodal points NNP is selected as 17, number of elements NEL=NNP − 1 or 16, and the nodal point at the bottom of the footing is NBX = (D/DX)+1 = 3/0.5 + 1 = 7. A schematic diagram of this problem is illustrated in Figure 5-6. The input parameters of data file VDIN.DAT for the consolidation/swell model, NOPT = 0, are given in Table F-4a. The maximum past pressures were omitted, which caused the program to assume these values equal with the swell pressures. The output data listed in data file VDOU.DAT shown in Table F-4b indicate substantial potential heave of 0.3 ft or 3.6 in. beneath the footing.

B. SETTLEMENT OF GRANULAR SOIL.
The same footing illustrated in Figure 5-6 is to be con-

structed in granular soil consisting of two distinctive layers. Field tests consisting of cone penetration and dilatometer data were obtained.

(1) The input parameters of data file VDIN.DAT for the Leonard and Frost model, NOPT = 1, are given in Table F-5a. The output data listed in data file VDOU.DAT shown in Table F-5b indicate 0.0092 ft or 0.11 in. of settlement beneath the footing.

(2) The same problem was applied to the Schmertmann model using the cone penetration resistance of the soil layers. The input data are given in Table F-6a. The output data shown in Table F-6b indicate 0.012 ft or 0.15 in. of settlement.

Table F-6. Settlement of Granular Soil, Schmertmann Model

a. Input Data, File VDIN.DAT

```
     FOOTING IN GRANULAR SOIL − SCHMERTMANN
 1     2    1    17     7     2      0.50
 1     1
12     2
16     2
 1     2.700     20.000     1.540
 2     2.65      19.300     0.900
    8.00     0     1
    1.00        3.00      3.00      0
 1    70.00
 2   100.00
 3    10.00
```

b. Output Data, File VDOU.DAT

```
     FOOTING IN GRANULAR SOIL − SCHMERTMANN

NUMBER OF PROBLEMS=    1      NUMBER OF NODAL POINTS=   17
NUMBER OF NODAL POINT AT BOTTOM OF FOUNDATION=        7
NUMBER OF DIFFERENT SOIL LAYERS=    2     INCREMENT DEPTH=      0.50   FT

     SCHMERTMANN MODEL

     RECTANGULAR SLAB FOUNDATION

DEPTH OF FOUNDATION =                3.00   FEET
TOTAL DEPTH OF THE SOIL PROFILE =    8.00   FEET
ELEMENT     NUMBER OF SOIL

    1          1
    2          1
    3          1
    4          1
    5          1
```

Table F-6. Continued

b. Output Data, File VDOU.DAT

ELEMENT	NUMBER OF SOIL
6	1
7	1
8	1
9	1
10	1
11	1
12	2
13	2
14	2
15	2
16	2

MATERIAL	SPECIFIC GRAVITY	WATER CONTENT, %	VOID RATIO
1	2.700	20.000	1.540
2	2.650	19.300	0.900

DEPTH TO WATER TABLE =   8.00   FEET

DISPLACEMENTS AT EACH DEPTH OUTPUT

EQUILIBRIUM SATURATED PROFILE ABOVE WATER TABLE

APPLIED PRESSURE ON FOUNDATION=   1.00 TSF
LENGTH =   3.00   FEET   WIDTH =   3.00   FEET

         CENTER OF FOUNDATION

MATERIAL	CONE RESISTANCE, TSF
1	70.00
2	100.00

TIME AFTER CONSTRUCTION IN YEARS=   10.00

ELEMENT	DEPTH, FT	SETTLEMENT, FT
7	3.25	−0.00069
8	3.75	−0.00138
9	4.25	−0.00205
10	4.75	−0.00222
11	5.25	−0.00193
12	5.75	−0.00115
13	6.25	−0.00096
14	6.75	−0.00078
15	7.25	−0.00060
16	7.75	−0.00042

SETTLEMENT BENEATH FOUNDATION=   −0.01218   FEET

Table F-7. Collapse Potential

a. Input Data, File VDIN.DAT

```
      FOOTING IN GRANULAR SOIL - SCHMERTMANN
   1    3    1   17    7    2       0.50
   1    1
  12    2
  16    2
   1    2.700    20.000    1.540
   2    2.65     19.300    0.900
      8.00    0    1
      1.00       3.00       3.00    0
   1      0.01      0.40     1.00     1.00     4.00
   2      0.05      0.40     1.00     1.00     4.00
   1      0.00      1.00     2.00    10.00    15.00
   2      0.00      0.80     1.50     8.00    12.00
```

b. Output Data, File VDOU.DAT

```
      FOOTING IN GRANULAR SOIL - SCHMERTMANN

NUMBER OF PROBLEMS=    1       NUMBER OF NODAL POINTS=    17
NUMBER OF NODAL POINT AT BOTTOM OF FOUNDATION=            7
NUMBER OF DIFFERENT SOIL LAYERS=    2       INCREMENT DEPTH=        0.50   FT

        COLLAPSIBLE SOIL

        RECTANGULAR SLAB FOUNDATION

DEPTH OF FOUNDATION =                   3.00   FEET
TOTAL DEPTH OF THE SOIL PROFILE =      8.00   FEET
ELEMENT      NUMBER OF SOIL
   1             1
   2             1
   3             1
   4             1
   5             1
   6             1
   7             1
   8             1
   9             1
  10             1
  11             1
  12             2
  13             2
  14             2
  15             2
  16             2
```

MATERIAL	SPECIFIC GRAVITY	WATER CONTENT, %	VOID RATIO
1	2.700	20.000	1.540
2	2.650	19.300	0.900

```
DEPTH TO WATER TABLE =               8.00   FEET

DISPLACEMENTS AT EACH DEPTH OUTPUT
```

Table F-7. Continued

b. Output Data, File VDOU.DAT

EQUILIBRIUM SATURATED PROFILE ABOVE WATER TABLE

APPLIED PRESSURE ON FOUNDATION=   1.00 TSF
LENGTH =   3.00 FEET   WIDTH =   3.00   FEET

      CENTER OF FOUNDATION

    APPLIED PRESSURE AT 5 POINTS IN UNITS OF TSF

MATERIAL	A	BB	B	C	D
1	0.01	0.40	1.00	1.00	4.00
2	0.05	0.40	1.00	1.00	4.00

    STRAIN AT 5 POINTS IN PERCENT

MATERIAL	A	BB	B	C	D
1	0.00	1.00	2.00	10.00	15.00
2	0.00	0.80	1.50	8.00	12.00

COLLAPSE DISTRIBUTION ABOVE FOUNDATION DEPTH

ELEMENT	DEPTH,FT	DELTA,FT
1	0.25	0.00007
2	0.75	−0.02081
3	1.25	−0.03052
4	1.75	−0.03691
5	2.25	−0.04169
6	2.75	−0.07354

COLLAPSE DISTRIBUTION BELOW FOUNDATION

ELEMENT	DEPTH,FT	DELTA,FT
7	3.25	−0.07999
8	3.75	−0.07955
9	4.25	−0.07834
10	4.75	−0.07674
11	5.25	−0.07516
12	5.75	−0.05423
13	6.25	−0.05269
14	6.75	−0.05171
15	7.25	−0.05119
16	7.75	−0.05104

SOIL COLLAPSE NEXT TO FOUNDATION EXCLUDING COLLAPSE
IN SUBSOIL BENEATH FOUNDATION = −0.10171 FEET
SUBSOIL COLLAPSE = −0.32532 FEET
TOTAL COLLAPSE = −0.42702 FEET

Table F-8. Listing of Program VDISPL

```
C     PREDICTION OF VERTICAL MOVEMENT, PROGRAM VDISPL.FOR
C     DEVELOPED BY L. D. JOHNSON
C     INPUT PARAMETERS
C       1ST LINE: NAME OF PROBLEM                           (20A4)
C       2ND LINE: NPROB NOPT NBPRES NNP NBX NMAT DX         (6I5,F10.2)
C       3RD LINE: N  IE(N,M)                                (2I5)
C     3RD LINE REPEATED FOR EACH NEW MATERIAL; LAST LINE IS NEL    NMAT
C       4TH LINE: M  G(M)  WC(M)  EO(M)                     (I5,3F10.3)
C     4TH LINE REPEATED FOR EACH NEW LAYER  M  UNTIL  M=NMAT
C       5TH LINE: DGWT  IOPTION  NOUT                       (F10.2,2I5)
C       6TH LINE: Q,BLEN,BWID,MRECT                         (3F10.2,I5)
C       7TH LINE: IF(NOPT.EQ.0) M SP(M) CS(M) CC(M) PM(M)   (I5,4F10.2)
C       7TH LINE: IF(NOPT.EQ.1) M PO(M) P1(M) QC(M)         (I5,3F10.2)
C       7TH LINE: IF(NOPT.EQ.2 OR 4) M QC(M)                (I5,F10.2)
C       7TH LINE: IF(NOPT.EQ.3) M PRES(M,J),J=1,5           (I5,5F10.2)
C     7TH LINE REPEATED FOR EACH DIFFERENT MATERIAL M UNTIL M=NMAT
C       8TH LINE: IF(NOPT.EQ.0) XA  XF                      (2F10.2)
C       8TH LINE: IF(NOPT.EQ.2.OR 4) TIME                   (F10.2)
C       8TH LINE: IF(NOPT.EQ.3) M STRA(M,J),J=1,5           (I5,5F10.2)
C     ABOVE LINE REPEATED FOR EACH DIFFERENT MATERIAL UNTIL M=NMAT
C     DESCRIPTION OF INPUT PARAMETERS
C         NAME OF PROBLEM  Insert title of your problem
C         NPROB    number of problems to solve with different active zone
C                  depths,groundwater level, moisture profile, and
C                  foundation dimensions
C         NOPT     Option for model, =0 for consolidation swell (MECH)
C                      =1 for Leonards and Frost
C                      =2 for Schmertmann
C                      =3 for Collapsible soil
C                      =4 for elastic soil settlement
C         NBPRES   Option for foundation, =1 for rectangular slab and
C                  =2 for long strip footing
C         NNP      Total number of nodal points
C         NBX      Number of nodal point at bottom of foundation
C         NMAT     Total number of different soil layers, < 10
C         DX       Increment of depth, ft
C         N        Element number
C         IE(N,1)  Number of soil layer  M  associated with element  N
C         M        Number of soil layer
C         G(M)     Specific gravity of soil layer  M
C         WC(M)    Water content of soil layer  M, Percent
C         EO(M)    Initial void ratio of soil layer  M
C         DGWT     Depth to hydrostatic water table, ft; If IOPTION =2,
C                  set DGWT=XA+UWA/0.03125 where UWA= suction (or
C                  positive value of the negative pore water pressure) at
C                  depth XA
C         IOPTION  Equilibrium moisture profile, = 0 for saturation above
C                  the water table; if NOPT=0, then =1 for hydrostatic
C                  with shallow water table, method 2 or =2 for
C                  hydrostatic without shallow water table
C         NOUT     Amount of output data, =0 only heave computations and
C                      =1  for heave and pore pressure at each depth
C                          increment
C         Q        Applied pressure on foundation, tsf
C         BLEN     Length of foundation or 0.0 if NBPRES = 2, feet
```

Table F-8. Continued

```
C           BWID      Width of foundation or long continuous footing, feet
C           MRECT     Location of calculation, =0 for center and =1
C                     for corner of rectangle or edge of long footing
C           SP(M)     Swell pressure of soil layer  M, tsf
C           CS(M)     Swell index of soil layer  M
C           CC(M)     Compression index of soil layer  M
C           PM(M)     Maximum past pressure, tsf
C           PO(M)     Dilatometer A pressure, tsf
C           P1(M)     Dilatometer B pressure, tsf
C           QC(M)     If NOPT=2, cone penetration resistance, tsf
C                     If NOPT=4, elastic soil modulus, tsf
C           PRES(M) Applied pressure at 5 points from collapsible soil,tsf
C           STRA(M) Strain at 5 points from collapsible soil
C           XA        Depth of active zone of heave, feet
C           XF        Depth from ground surface to the depth that the active
C                     zone begins, feet
C           TIME      Time in years after Construction for Schmertmann model
      PARAMETER (NL=30,NQ=101)
      COMMON/SLA/P(NQ),IE(NQ,1),EO(NL),DX,NBX,NEL,PII,NOUT
      DIMENSION PP(NQ),G(NL),WC(NL),HED(20)
      OPEN(5,FILE='VDIN.DAT')
      OPEN(6,FILE='VDOU.DAT')
      READ(5,1) (HED(I),I=1,20)
      WRITE(6,1) (HED(I),I=1,20)
    1 FORMAT(20A4)
      GAW=0.03125
      PII=3.14159265
      NP=1
      READ(5,2) NPROB,NOPT,NBPRES,NNP,NBX,NMAT,DX
    2 FORMAT(6I5,F10.2)
      WRITE(6,3) NPROB,NNP,NBX,NMAT,DX
    3 FORMAT(/,1X,'NUMBER OF PROBLEMS=',I5,5X,'NUMBER OF NODAL POINTS='
     1,I5,/,1X,'NUMBER OF NODAL POINT AT BOTTOM OF FOUNDATION=',I11,/,1X
     2,'NUMBER OF DIFFERENT SOIL LAYERS=',I5,5X,'INCREMENT DEPTH=',F10.2
     3,'  FT',/)
      IF(NOPT.EQ.0)WRITE(6,4)
    4 FORMAT(10X,'CONSOLIDATION SWELL MODEL',/)
      IF(NOPT.EQ.1)WRITE(6,5)
    5 FORMAT(10X,'LEONARDS AND FROST MODEL',/)
      IF(NOPT.EQ.2)WRITE(6,6)
    6 FORMAT(10X,'SCHMERTMANN MODEL',/)
      IF(NOPT.EQ.3)WRITE(6,7)
    7 FORMAT(10X,'COLLAPSIBLE SOIL',/)
      IF(NOPT.EQ.4)WRITE(6,8)
    8 FORMAT(10X,'ELASTIC SOIL',/)
      IF(NBPRES.EQ.1)WRITE(6,9)
    9 FORMAT(10X,'RECTANGULAR SLAB FOUNDATION',/)
      IF(NBPRES.EQ.2)WRITE(6,10)
   10 FORMAT(10X,'LONG CONTINUOUS STRIP FOUNDATION',/)
      DEPF=DX*FLOAT(NBX-1)
      WRITE(6,11)DEPF
   11 FORMAT(1X,'DEPTH OF FOUNDATION =',12X,F10.2,'  FEET')
      DEPPR = DX*FLOAT(NNP-1)
      WRITE(6,12)DEPPR
```

Table F-8. Continued

```
  12 FORMAT(1X,'TOTAL DEPTH OF THE SOIL PROFILE =',F10.2,'  FEET')
     NEL=NNP-1
     L=0
     WRITE(6,21)
  21 FORMAT(1X,'ELEMENT     NUMBER OF SOIL',/)
  22 READ(5,2)N,IE(N,1)
  25 L=L+1
     IF(N-L)35,35,30
  30 IE(L,1)=IE(L-1,1)
     WRITE(6,32)L,IE(L,1)
  32 FORMAT(I5,8X,I5)
     GOTO 25
  35 WRITE(6,32)L,IE(L,1)
     IF(NEL-L)40,40,22
  40 CONTINUE
     WRITE(6,390)
 390 FORMAT(/,1X,'MATERIAL    SPECIFIC GRAVITY   WATER CONTENT, %   ',
    1'VOID RATIO',/)
 400 READ(5,401) M,G(M),WC(M),EO(M)
 401 FORMAT(I5,3F10.3)
     IF(NMAT-M)403,405,400
 403 WRITE(6,404) M
 404 FORMAT(/,5X,'ERROR IN MATERIAL', I5)
     STOP
 405 DO 410 M=1,NMAT
     WRITE(6,407) M,G(M),WC(M),EO(M)
 407 FORMAT(I5,3F18.3)
 410 CONTINUE
1000 READ(5,45) DGWT,IOPTION,NOUT
  45 FORMAT(F10.2,2I5)
     READ(5,46)Q,BLEN,BWID,MRECT
  46 FORMAT(3F10.2,I5)
     WRITE(6,50) DGWT
  50 FORMAT(1X,'DEPTH TO WATER TABLE =',11X,F10.2,'  FEET',/)
     IF(NOUT.EQ.1)WRITE(6,51)
     IF(NOUT.EQ.0)WRITE(6,52)
  51 FORMAT(1X,'DISPLACEMENTS AT EACH DEPTH OUTPUT',/)
  52 FORMAT(1X,'TOTAL DISPLACEMENTS ONLY',/)
     IF(IOPTION.EQ.0.OR.NOPT.EQ.1)WRITE(6,61)
  61 FORMAT(1X,'EQUILIBRIUM SATURATED PROFILE ABOVE WATER TABLE',/)
     IF(IOPTION.EQ.1.AND.NOPT.EQ.0)WRITE(6,62)
  62 FORMAT(1X,'EQUILIBRIUM HYDROSTATIC PROFILE ABOVE WATER TABLE',/)
     WRITE(6,90)Q,BLEN,BWID
  90 FORMAT(/,1X,'APPLIED PRESSURE ON FOUNDATION=',F10.2,' TSF',/,1X,
    1'LENGTH =',F10.2,'  FEET',5X,'WIDTH =',F10.2,'  FEET',/)
     IF(MRECT.EQ.0)WRITE(6,91)
  91 FORMAT(9X,'CENTER OF FOUNDATION',/)
     IF(MRECT.EQ.1)WRITE(6,92)
  92 FORMAT(9X,'CORNER OF SLAB OR EDGE OF LONG STRIP FOOTING',/)
C            CALCULATION OF EFFECTIVE OVERBURDEN PRESSURE
 105 P(1)=0.0
     PP(1)=0.0
     DXX=DX
     DO 110 I=2,NNP
     MTYP=IE(I-1,1)
```

Table F-8. Continued

```
      WCC=WC(MTYP)/100.
      GAMM=G(MTYP)*GAW*(1.+WCC)/(1.+EO(MTYP))
      IF(DXX.GT.DGWT)GAMM=GAMM-GAW
      P(I)=P(I-1)+DX*GAMM
      PP(I)=P(I)
      DXX=DXX+DX
  110 CONTINUE
      IF(NOPT.NE.0.OR.IOPTION.EQ.0)GOTO 120
      MO=IFIX(DGWT/DX)
      IF(MO.GT.NNP)MO=NNP
      DO 117 I=1,MO
      BN=DGWT/DX-FLOAT(I-1)
      P(I)=P(I)+BN*DX*GAW
  117 CONTINUE
  120 CALL SLAB(Q,BLEN,BWID,MRECT,NBPRES,PP(NBX))
C           CALCULATION OF MOVEMENT FROM MODELS
      IF(NOPT.EQ.0) CALL MECH(NMAT)
      IF(NOPT.EQ.1) CALL LEON(Q,NMAT,DGWT,BWID,PP,NBPRES)
      IF(NOPT.EQ.2) CALL SCHMERT(Q,NMAT,DGWT,BWID,PP,NBPRES,2)
      IF(NOPT.EQ.3) CALL COLL(NMAT)
      IF(NOPT.EQ.4) CALL SCHMERT(Q,NMAT,DGWT,BWID,PP,NBPRES,4)
      NP=NP+1
      IF(NP.GT.NPROB) GOTO 200
      GOTO 1000
  200 CLOSE(5,STATUS='KEEP')
      CLOSE(6,STATUS='KEEP')
      STOP
      END
C
C
      SUBROUTINE MECH(NMAT)
      PARAMETER(NL=30,NQ=101)
      COMMON/SLA/P(NQ),IE(NQ,1),EO(NL),DX,NBX,NEL,PII,NOUT
      DIMENSION SP(NL),CS(NL),CC(NL),PM(NL)
      WRITE(6,5)
    5 FORMAT(/,1X,'MATERIAL   SWELL PRESSURE,     SWELL          COMPRESSION
     1    MAXIMUM PAST',/,1X,'                        TSF            INDEX
     2  INDEX        PRESSURE,TSF',/)
      DO 10 I = 1,NMAT
      READ(5,11) M,SP(M),CS(M),CC(M),PM(M)
      IF(PM(M).LT.SP(M)) WRITE(6,14) SP(M),PM(M)
      IF(PM(M).LT.SP(M)) PM(M)=SP(M)
      WRITE(6,24)M,SP(M),CS(M),CC(M),PM(M)
   10 CONTINUE
   11 FORMAT(I5,4F10.4)
   14 FORMAT(/,1X,'SWELL PRESSURE',F10.2,'  WAS SET GREATER THAN MAXIM',
     1'UM PAST PRESSURE',F10.2,/,1X,'WHICH IS NOT POSSIBLE; SWELL PRESSU
     2RE SET EQUAL TO MAXIMUM PAST PRESSURE',/)
   24 FORMAT(1X,I5,4F15.3)
C
      READ(5,30)XA,XF
   30 FORMAT(2F10.2)
      WRITE(6,31) XA,XF
```

Table F-8. Continued

```
 31 FORMAT(/,8X,'ACTIVE ZONE DEPTH (FT) =',F10.2,/,1X,'DEPTH ACTIVE ZO
   1NE BEGINS (FT) =',F10.2,/)
    DELH1=0.0
    DXX=0.0
    CALL PSAD(N1,N2,XA,XF,DXX,DX,NBX)
    IF(N1.GE.N2) GOTO 50
    IF(NOUT.EQ.0) GOTO 50
    WRITE(6,32)
 32 FORMAT(/,1X,'HEAVE DISTRIBUTION ABOVE FOUNDATION DEPTH',/,1X,'ELEM
   1ENT    DEPTH,FT    DELTA HEAVE,FT    EXCESS PORE PRESSURE,TSF',/)
    DO 40 I=N1,N2
    MTYP=IE(I,1)
    PR=(P(I)+P(I+1))/2.
    CA=SP(MTYP)/PR
    CB=SP(MTYP)/PM(MTYP)
    CBB=PM(MTYP)/PR
    E=EO(MTYP)+CS(MTYP)*ALOG10(CA)
    IF(PR.GT.PM(MTYP)) E=EO(MTYP)+CS(MTYP)*ALOG10(CB)+CC(MTYP)*ALOG10
   1(CBB)
    DEL=(E-EO(MTYP))/(1.+EO(MTYP))
    IF(NOUT.EQ.0) GOTO 36
    DELP=SP(MTYP)-PR
    WRITE(6,110) I,DXX,DEL,DELP
 36 DELH1=DELH1+DX*DEL
    DXX=DXX+DX
 40 CONTINUE
 50 DELH2=0.0
    IF(NBX.GT.NEL) GOTO 120
    DXX=FLOAT(NBX)*DX-DX/2.
    IF(NOUT.EQ.0) GOTO 65
    WRITE(6,60)
 60 FORMAT(/,1X,'HEAVE DISTRIBUTION BELOW FOUNDATION',/,1X,'ELEMENT
   1 DEPTH,FT    DELTA HEAVE,FT    EXCESS PORE PRESSURE,TSF',/)
 65 DO 100 I=NBX,NEL
    MTYP=IE(I,1)
    PR=(P(I)+P(I+1))/2.
    CA=SP(MTYP)/PR
    CB=SP(MTYP)/PM(MTYP)
    CBB=PM(MTYP)/PR
    E=EO(MTYP)+CS(MTYP)*ALOG10(CA)
    IF(PR.GT.PM(MTYP))E=EO(MTYP)+CS(MTYP)*ALOG10(CB)+CC(MTYP)*ALOG10
   1(CBB)
    DEL=(E-EO(MTYP))/(1.+EO(MTYP))
    IF(NOUT.EQ.0) GOTO 80
    DELP=SP(MTYP)-PR
    WRITE(6,110) I,DXX,DEL,DELP
 80 DELH2=DELH2+DX*DEL
    DXX=DXX+DX
100 CONTINUE
110 FORMAT(I5,F13.2,F18.5,5X,F18.5)
    DEL1=DELH1+DELH2
    WRITE(6,305) DELH1,DELH2,DEL1
305 FORMAT(/,1X,'SOIL HEAVE NEXT TO FOUNDATION EXCLUDING HEAVE',/,1X,
   1'IN SUBSOIL BENEATH FOUNDATION =',F8.5,'  FEET',//,14X,'SUBSOIL ',
   2'MOVEMENT =',F8.5,'  FEET',/,19X,'TOTAL HEAVE =',F8.5,'  FEET')
```

Table F-8. Continued

```
    120 RETURN
        END
C
C
        SUBROUTINE SLAB(Q,BLEN,BWID,MRECT,NBPRES,WT)
        PARAMETER (NL=30,NQ=101)
        COMMON/SLA/P(NQ),IE(NQ,1),EO(NL),DX,NBX,NEL,PII,NOUT
C
C       CALCULATION OF SURCHARGE PRESSURE FROM STRUCTURE
C
        NNP=NEL+1
        ANBX=FLOAT(NBX)*DX
        DXX=0.0
        BPRE1=Q-WT
        BPRES=BPRE1
        DO 100 I=NBX,NNP
        IF(DXX.LT.0.01) GOTO 80
        MTYP=IE(I-1,1)
        IF(NBPRES.EQ.2) GOTO 70
        BL=BLEN
        BW=BWID
        BPR=BPRES
        IF(MRECT.EQ.1) GOTO 50
        BL=BLEN/2.
        BW=BWID/2.
     50 VE2=(BL**2.+BW**2.+DXX**2.)/(DXX**2.)
        VE=VE2**0.5
        AN=BL*BW/(DXX**2.)
        AN2=AN**2.
        ENM=(2.*AN*VE/(VE2+AN2))*(VE2+1.)/VE2
        FNM=2.*AN*VE/(VE2-AN2)
        IF(MRECT.EQ.1)BPR=BPRES/4.
        AB=ATAN(FNM)
        IF(FNM.LT.0.) AB=PII+AB
        P(I)=P(I)+BPR*(ENM+AB)/PII
        GOTO 90
     70 DB=DXX/BWID
        PS=-0.157-0.22*DB
        IF(MRECT.EQ.0.AND.DB.LT.2.5)PS=-0.28*DB
        PS=10.**PS
        P(I)=P(I)+BPRES*PS
        GOTO 90
     80 P(I)=P(I)+BPRES
     90 DXX=DXX+DX
    100 CONTINUE
        RETURN
        END
C
C
C
        SUBROUTINE PSAD(N1,N2,XA,XF,DXX,DX,NBX)
        AN1=XF/DX
        AN2=XA/DX
        N1=IFIX(AN1)+1
        N2=AN2
```

Table F-8. Continued

```
      DXX=XF+DX/2.
      N3=NBX-1
      IF(N2.GT.N3)N2=N3
      CONTINUE
      RETURN
      END
C
C
      SUBROUTINE LEON(Q,NMAT,DGWT,BWID,PP,NBPRES)
      PARAMETER(NL=30,NQ=101)
      COMMON/SLA/P(NQ),IE(NQ,1),EO(NL),DX,NBX,NEL,PII,NOUT
      DIMENSION PO(NL),P1(NL),QC(NL),PP(NQ)
      WRITE(6,5)
    5 FORMAT(/,1X,'MATERIAL   A PRESSURE, TSF   B PRESSURE, TSF   CONE
     1RESISTANCE, TSF',/)
      DO 10 I = 1,NMAT
      READ(5,15)M,PO(M),P1(M),QC(M)
      WRITE(6,20)M,PO(M),P1(M),QC(M)
   10 CONTINUE
   15 FORMAT(I5,3F10.2)
   20 FORMAT(I5,3F18.2)
      CALL SPLINE
      NNP=NEL+1
      GAW=0.03125
      DELH=0.0
      DEL=0.0
      QNET=Q-PP(NBX)
      WRITE(6,17)QNET
   17 FORMAT(/,1X,'QNET=',F10.5)
      DXX=DX*FLOAT(NBX) - DX/2.
      C1=1 - 0.5*PP(NBX)/QNET
      IF(C1.LT.0.5) C1=0.5
      IF(NOUT.EQ.0) GOTO 30
      WRITE(6,25)
   25 FORMAT(/,1X,'ELEMENT   DEPTH,    SETTLEMENT,      KO    QC/SIGV PHI,
     1 DEGREES',/,1X,'              FT           FT',/)
   30 DO 300 I=NBX,NEL
      MTYP=IE(I,1)
      PR1=(PP(I+1)+PP(I))/2.
      PR=(P(I+1)+P(I))/2.
      UW=0.0
      IF(DXX.GT.DGWT) UW=(DXX-DGWT)*GAW
      AKD = (PO(MTYP)-UW)/PR1
      AID = (P1(MTYP)-PO(MTYP))/(PO(MTYP)-UW)
      ED = 34.7*(P1(MTYP)-PO(MTYP))
      RQC = QC(MTYP)/PR1
      AKO=0.376+0.095*AKD-0.0017*RQC
      EC = AKO
      S3 = RQC
      MM=0
      UC=0.0
      CALL BICUBE(UC,EC,S3)
      S3=RQC
      PHIPS=UC*PII/180.
      AKA = (1-SIN(PHIPS))/(1+SIN(PHIPS))
```

Table F-8. Continued

```
      AKP = (1+SIN(PHIPS))/(1-SIN(PHIPS))
      AAK = AKO
      IF(AKO.LE.AKA)AAK=AKA
      IF(AKO.GE.AKP)AAK=AKP
      IF(ABS(AAK-AKO).GT.0.01)EC=AAK
      IF(ABS(AAK-AKO).GT.0.01) CALL BICUBE(UC,EC,S3)
      S3=RQC
      PHIAX=UC
      IF(UC.GT.32.0) PHIAX=UC-((UC-32.)/3.)
      PHI=PHIAX*PII/180.
      OCR=(AAK/(1-SIN(PHI)))**(1./(0.8*SIN(PHI)))
      PM=OCR*PR1
      ROC=(PM-PR1)/(PR-PR1)
      IF(ROC.LT.0.0)ROC=0.0
      RNC=(PR-PM)/(PR-PR1)
      IF(RNC.LT.0.0)RNC=0.0
      IF(NBPRES.EQ.2) GOTO 100
      ANN=0.5*BWID/DX + DX*FLOAT(NBX-1)
      NN=IFIX(ANN)
      SIGM=PR1
      AIZP=0.5+0.1*(QNET/SIGM)**0.5
      DEPT=DXX-FLOAT(NBX-1)*DX
      AIZ=0.1+(AIZP-0.1)*DEPT/(0.5*BWID)
      IF(DEPT.GT.0.5*BWID)AIZ=AIZP+AIZP/3.0-AIZP*DEPT/(1.5*BWID)
      IF(DEPT.GT.2*BWID)AIZ=0.0
      GOTO 200
  100 ANN=BWID/DX + DX*FLOAT(NBX-1)
      NN=IFIX(ANN)
      SIGM=PR1
      AIZP=0.5+0.1*(QNET/SIGM)**0.5
      DEPT=DXX-FLOAT(NBX-1)*DX
      AIZ=0.2+(AIZP-0.2)*DEPT/BWID
      IF(DEPT.GT.BWID)AIZ=AIZP+AIZP/3.-AIZP*DEPT/(3.*BWID)
      IF(DEPT.GT.4*BWID)AIZ=0.0
  200 F=0.7
      IF(ROC.LE.0.0)F=0.9
      DEL=-C1*QNET*AIZ*DX*(ROC/(3.5*ED)+RNC/(F*ED))
      DELH=DELH+DEL
      IF(NOUT.EQ.1)WRITE(6,310)I,DXX,DEL,EC,S3,UC
      DXX=DXX+DX
  300 CONTINUE
  310 FORMAT(I5,F10.2,F13.5,3F10.2)
      WRITE(6,320) DELH
  320 FORMAT(/,1X,'SETTLEMENT BENEATH FOUNDATION=',F10.5,'  FEET',/)
      RETURN
      END
C
C
      BLOCK DATA
      DIMENSION XX(100),YY(100),U(100)
      COMMON/SPL/XX,YY,U
      DATA(XX(I),I=1,99,11)/9*10./,(XX(I),I=2,99,11)/9*20./,(XX(I),I=3,
     199,11)/9*30./,(XX(I),I=4,99,11)/9*50./,(XX(I),I=5,99,11)/9*100./,
     2(XX(I),I=6,99,11)/9*200./,(XX(I),I=7,99,11)/9*300./,(XX(I),I=8,99,
```

Table F-8. Continued

```
    311)/9*500./,(XX(I),I=9,99,11)/9*1000./,(XX(I),I=10,99,11)/9*2000./
    4,(XX(I),I=11,99,11)/9*3000./
     DATA(YY(I),I=1,11)/11*0.16/,(YY(I),I=12,22)/11*0.20/,(YY(I),I=23,
    133)/11*0.4/,(YY(I),I=34,44)/11*0.6/,(YY(I),I=45,55)/11*0.8/,(YY(I)
    2,I=56,66)/11*1.0/,(YY(I),I=67,77)/11*2.0/,(YY(I),I=78,88)/11*4./,
    3(YY(I),I=89,99)/11*6./
     DATA(U(I),I=1,99)/25.,30.1,33.2,36.4,39.9,42.8,44.4,46.,48.5,50.5,
    152.,24.8,30.,33.,36.2,39.7,42.6,44.2,45.8,48.2,50.2,51.5,24.5,29.7
    2,32.6,35.6,39.3,42.1,43.7,45.4,47.5,49.7,51.,24.2,29.5,32.2,35.1,
    338.8,41.7,43.3,45.,47.2,49.,50.,24.,29.2,31.7,34.7,38.4,41.4,42.9,
    444.6,46.8,48.6,49.7,23.8,28.8,31.5,34.4,38.,41.,42.5,44.3,46.5,48.
    54,49.5,23.,27.5,30.,33.,36.6,39.6,41.2,43.,45.4,47.2,48.4,22.,26.,
    628.3,31.2,34.5,37.7,39.7,41.5,43.7,45.7,47.,21.,25.,27.2,30.,33.8,
    736.,38.2,40.3,42.7,44.8,46.1/
     END
C
C

     SUBROUTINE SPLINE
C    SPLINE TO CALCULATE VARIABLES
     DIMENSION XX(100),YY(100),U(100)
     COMMON/SPL/XX,YY,U
     COMMON/SPLIN/X(100),Y(100),S(100)
     COMMON/BICUB/UX(100),UY(100),UXY(100)
     NCONF=11
     NSTR=9
     NCT=1
     N=NCONF
 210 ID=NCONF*NCT
     II=ID-NCONF+1
     JJ=NCONF-1
     DO 220 I=II,ID
     J=NCONF-JJ
     X(J)=XX(I)
     Y(J)=U(I)
 220 JJ=JJ-1
     CALL SOLV(N)
     IC=1
     DO 225 I=II,ID
     UX(I)=S(IC)
 225 IC=IC+1
     NCT=NCT+1
     IF(NCT.LE.NSTR)GO TO 210
     NCT=1
     N=NSTR
 230 IT=NCONF*(NSTR-1)
     ID=IT+NCT
     II=ID-IT
     JJ=NSTR-1
     DO 235 I=II,ID,NCONF
     J=NSTR-JJ
     X(J)=YY(I)
     Y(J)=U(I)
 235 JJ=JJ-1
     CALL SOLV(N)
     IC=1
```

Table F-8. Continued

```
      DO 240 I=II,ID,NCONF
      UY(I)=S(IC)
  240 IC=IC+1
      NCT=NCT+1
      IF(NCT.LE.NCONF)GO TO 230
      NCT=1
      N=NCONF
  243 ID=NCONF*NCT
      II=ID-NCONF+1
      JJ=NCONF-1
      DO 245 I=II,ID
      J=NCONF-JJ
      X(J)=XX(I)
      Y(J)=UY(I)
  245 JJ=JJ-1
      CALL SOLV(N)
      IC=1
      DO 250 I=II,ID
      UXY(I)=S(IC)
  250 IC=IC+1
  300 FORMAT(I5,F15.5)
      NCT=NCT+1
      IF(NCT.LE.NSTR)GO TO 243
      RETURN
      END
C
C
C
      SUBROUTINE SOLV(N)
      COMMON/SPLIN/X(100),Y(100),S(100)
      DIMENSION A(100),B(100),C(100),D(100),F(100),GG(100),H(100)
      N1=N-1
      DO 2010 I=2,N
 2010 H(I)=X(I)-X(I-1)
      DO 2020 I=2,N
 2020 A(I)=1./H(I)
      A(1)=0.
      DO 2030 I=2,N1
      T1=1./H(I)+1./H(I+1)
 2030 B(I)=2.*T1
      B(1)=2.*(1./H(2))
      B(N)=2.*(1./H(N))
      DO 2040 I=1,N1
 2040 C(I)=1./H(I+1)
      C(N)=0.
      DO 2050 I=2,N1
      T1=(Y(I)-Y(I-1))/(H(I)*H(I))
      T2=(Y(I+1)-Y(I))/(H(I+1)*H(I+1))
 2050 D(I)=3.*(T1+T2)
      T1=(Y(2)-Y(1))/(H(2)*H(2))
      D(1)=3.*T1
      T2=(Y(N)-Y(N-1))/(H(N)*H(N))
      D(N)=3.*T2
C     FORWARD PASS
      GG(1)=C(1)/B(1)
```

Table F-8. Continued

```
       DO 2100 I=2,N1
       T1=B(I)-A(I)*GG(I-1)
 2100  GG(I)=C(I)/T1
       F(1)=D(1)/B(1)
       DO 2110 I=2,N
       TEM=D(I)-A(I)*F(I-1)
       T1=B(I)-A(I)*GG(I-1)
 2110  F(I)=TEM/T1
C      BACK SOLUTION
       S(N)=F(N)
       I=N-1
 2120  S(I)=F(I)-GG(I)*S(I+1)
       IF(I.EQ.1) GOTO 2150
       I=I-1
       GO TO 2120
 2150  CONTINUE
       RETURN
       END
C
C
       SUBROUTINE BICUBE(UC,EC,S3)
       DIMENSION XX(100),YY(100),U(100)
       COMMON/SPL/XX,YY,U
       COMMON/BICUB/UX(100),UY(100),UXY(100)
       DIMENSION H(16),KE(100,4)
       DATA((KE(K,I),I=1,4),K=1,10)/1,2,13,12,2,3,14,13,3,4,15,14,4,5,16,
      115,5,6,17,16,6,7,18,17,7,8,19,18,8,9,20,19,9,10,21,20,10,11,22,21/
       DATA((KE(K,I),I=1,4),K=11,20)/12,13,24,23,13,14,25,24,14,15,26,25,
      115,16,27,26,16,17,28,27,17,18,29,28,18,19,30,29,19,20,31,30,20,21,
      232,31,21,22,33,32/
       DATA((KE(K,I),I=1,4),K=21,30)/23,24,35,34,24,25,36,35,25,26,37,36,
      126,27,38,37,27,28,39,38,28,29,40,39,29,30,41,40,30,31,42,41,31,32,
      243,42,32,33,44,43/
       DATA((KE(K,I),I=1,4),K=31,40)/34,35,46,45,35,36,47,46,36,37,48,47,
      137,38,49,48,38,39,50,49,39,40,51,50,40,41,52,51,41,42,53,52,42,43,
      254,53,43,44,55,54/
       DATA((KE(K,I),I=1,4),K=41,50)/45,46,57,56,46,47,58,57,47,48,59,58,
      148,49,60,59,49,50,61,60,50,51,62,61,51,52,63,62,52,53,64,63,53,54,
      265,64,54,55,66,65/
       DATA((KE(K,I),I=1,4),K=51,60)/56,57,68,67,57,58,69,68,58,59,70,69,
      159,60,71,70,60,61,72,71,61,62,73,72,62,63,74,73,63,64,75,74,64,65,
      276,75,65,66,77,76/
       DATA((KE(K,I),I=1,4),K=61,70)/67,68,79,78,68,69,80,79,69,70,81,80,
      170,71,82,81,71,72,83,82,72,73,84,83,73,74,85,84,74,75,86,85,75,76,
      287,86,76,77,88,87/
       DATA((KE(K,I),I=1,4),K=71,80)/78,79,90,89,79,80,91,90,80,81,92,91,
      181,82,93,92,82,83,94,93,83,84,95,94,84,85,96,95,85,86,97,96,86,87,
      298,97,87,88,99,98/
       DO 400 M=1,80
       I=KE(M,1)
       J=KE(M,2)
       K=KE(M,3)
       L=KE(M,4)
       IF(S3.GE.XX(I).AND.S3.LE.XX(J))GOTO 410
       GOTO 400
```

Table F-8. Continued

```
  410 IF(EC.GE.YY(I).AND.EC.LE.YY(L))GOTO 420
  400 CONTINUE
  420 CONTINUE
      AA=XX(J)-XX(I)
      BB=YY(L)-YY(I)
      S3N=S3-XX(I)
      ECN=EC-YY(I)
      SL=S3N/AA
      T=ECN/BB
      S2=SL*SL
      S3=SL*SL*SL
      T2=T*T
      T3=T2*T
      F1=1.-3.*S2+2.*S3
      F2=S2*(3.-2.*SL)
      F3=AA*SL*(1.-SL)*(1.-SL)
      F4=AA*S2*(SL-1.)
      DO 430 KJ=1,2
      G1=1.-3.*T2+2.*T3
      G2=T2*(3.-2.*T)
      G3=BB*T*(1.-T)*(1.-T)
      G4=BB*T2-2.*T
      H(1)=F1*G1*U(I)
      H(2)=F2*G1*U(J)
      H(3)=F2*G2*U(K)
      H(4)=F1*G2*U(L)
      H(5)=F3*G1*UX(I)
      H(6)=F4*G1*UX(J)
      H(7)=F4*G2*UX(K)
      H(8)=F3*G2*UX(L)
      H(9)=F1*G3*UY(I)
      H(10)=F2*G3*UY(J)
      H(11)=F2*G4*UY(K)
      H(12)=F1*G4*UY(L)
      H(13)=F3*G3*UXY(I)
      H(14)=F4*G4*UXY(J)
      H(15)=F4*G4*UXY(K)
      H(16)=F3*G4*UXY(L)
      UC=0.0
      DO 480 KK=1,16
  480 UC=UC+H(KK)
  430 CONTINUE
      RETURN
      END
C
C
      SUBROUTINE SCHMERT(Q,NMAT,DGWT,BWID,PP,NBPRES,JOPT)
      PARAMETER(NL=30,NQ=101)
      COMMON/SLA/P(NQ),IE(NQ,1),EO(NL),DX,NBX,NEL,PII,NOUT
      DIMENSION QC(NL),PP(NQ)
      IF(JOPT.EQ.2)WRITE(6,5)
    5 FORMAT(/,1X,'MATERIAL    CONE RESISTANCE, TSF',/)
      IF(JOPT.EQ.4)WRITE(6,6)
    6 FORMAT(/,1X,'MATERIAL    ELASTIC MODULUS, TSF',/)
      DO 10 I = 1,NMAT
```

Table F-8. Continued

```
      READ(5,15)M,QC(M)
      WRITE(6,20)M,QC(M)
   10 CONTINUE
   15 FORMAT(I5,F10.2)
   20 FORMAT(I5,F18.2)
      READ(5,30)TIME
   30 FORMAT(F10.2)
      WRITE(6,35)TIME
   35 FORMAT(/,1X,'TIME AFTER CONSTRUCTION IN YEARS=',F10.2,/)
      NNP=NEL+1
      DELH=0.0
      DEL=0.0
      QNET=Q-PP(NBX)
      DXX=DX*FLOAT(NBX) - DX/2.
      C1=1 - 0.5*PP(NBX)/QNET
      IF(C1.LT.0.5) C1=0.5
      FF=TIME/0.1
      CT=1+0.2*ALOG10(FF)
      IF(NOUT.EQ.0) GOTO 40
      WRITE(6,25)
   25 FORMAT(/,1X,'ELEMENT    DEPTH, FT    SETTLEMENT, FT ',/)
   40 DO 300 I=NBX,NEL
      MTYP=IE(I,1)
      PR1=(PP(I+1)+PP(I))/2.
      ESI=QC(MTYP)
      IF(NBPRES.EQ.1.AND.JOPT.EQ.2)ESI=2.5*QC(MTYP)
      IF(NBPRES.EQ.2.AND.JOPT.EQ.2)ESI=3.5*QC(MTYP)
      IF(NBPRES.EQ.2) GOTO 100
      ANN=0.5*BWID/DX + DX*FLOAT(NBX-1)
      NN=IFIX(ANN)
      SIGM=PR1
      AIZP=0.5+0.1*(QNET/SIGM)**0.5
      DEPT=DXX-FLOAT(NBX-1)*DX
      AIZ=0.1+(AIZP-0.1)*DEPT/(0.5*BWID)
      IF(DEPT.GT.0.5*BWID)AIZ=AIZP+AIZP/3.0-AIZP*DEPT/(1.5*BWID)
      IF(DEPT.GT.2*BWID)AIZ=0.0
      GOTO 200
  100 ANN=BWID/DX + DX*FLOAT(NBX-1)
      NN=IFIX(ANN)
      SIGM=PR1
      AIZP=0.5+0.1*(QNET/SIGM)**0.5
      DEPT=DXX-FLOAT(NBX-1)*DX
      AIZ=0.2+(AIZP-0.2)*DEPT/BWID
      IF(DEPT.GT.BWID)AIZ=AIZP+AIZP/3.-AIZP*DEPT/(3.*BWID)
      IF(DEPT.GT.4*BWID)AIZ=0.0
  200 DEL=-C1*CT*QNET*AIZ*DX/ESI
      DELH=DELH+DEL
      IF(NOUT.EQ.1)WRITE(6,310)I,DXX,DEL
      DXX=DXX+DX
  300 CONTINUE
  310 FORMAT(I5,F13.2,F18.5)
      WRITE(6,320) DELH
  320 FORMAT(/,1X,'SETTLEMENT BENEATH FOUNDATION=',F10.5,'  FEET',/)
      RETURN
      END
```

Table F-8. Continued

```
C
C
      SUBROUTINE COLL(NMAT)
      PARAMETER(NL=30,NQ=101)
      COMMON/SLA/P(NQ),IE(NQ,1),EO(NL),DX,NBX,NEL,PII,NOUT
      DIMENSION PRES(NL,5),STRA(NL,5)
      WRITE(6,5)
    5 FORMAT(/,10X,'APPLIED PRESSURE AT 5 POINTS IN UNITS OF TSF',/,1X,'
     1MATERIAL     A          BB         B          C          D',/)
      DO 10 I = 1,NMAT
      READ(5,11) M,(PRES(M,J),J=1,5)
      WRITE(6,11)M,(PRES(M,J),J=1,5)
   10 CONTINUE
   11 FORMAT(I5,5F10.2)
      WRITE(6,15)
   15 FORMAT(/,10X,'STRAIN AT 5 POINTS IN PERCENT',/1X,'MATERIAL     A
     1      BB        B          C          D',/)
      DO 20 I=1,NMAT
      READ(5,21) M,(STRA(M,J),J=1,5)
      WRITE(6,21)M,(STRA(M,J),J=1,5)
      DO 19 K=1,5
      STRA(M,K) = STRA(M,K)/100.
   19 CONTINUE
   20 CONTINUE
   21 FORMAT(I5,5F10.2)
      DELH1=0.0
      DXX=DX/2.
      IF(NBX.EQ.1) GOTO 50
      IF(NOUT.EQ.0) GOTO 50
      WRITE(6,32)
   32 FORMAT(/,1X,'COLLAPSE DISTRIBUTION ABOVE FOUNDATION DEPTH',/,1X,'E
     1LEMENT     DEPTH,FT          DELTA,FT',/)
      DO 40 I=1,NBX-1
      MTY=IE(I,1)
      PR=(P(I)+P(I+1))/2.
      PRA=PRES(MTY,2)/PRES(MTY,1)
      PRB=PRES(MTY,3)/PRES(MTY,2)
      PRC=PRES(MTY,4)/PRES(MTY,1)
      PRD=PRES(MTY,5)/PRES(MTY,4)
      PRE=PRES(MTY,1)/PR
      PRF=PRES(MTY,2)/PR
      PRG=PRES(MTY,4)/PR
      SA=(STRA(MTY,2)-STRA(MTY,1))/ALOG10(PRA)
      SB=(STRA(MTY,3)-STRA(MTY,2))/ALOG10(PRB)
      SC=(STRA(MTY,4)-STRA(MTY,1))/ALOG10(PRC)
      SD=(STRA(MTY,5)-STRA(MTY,4))/ALOG10(PRD)
      IF(PR.LE.PRES(MTY,2))DEB=-STRA(MTY,1)+SA*ALOG10(PRE)
      IF(PR.GT.PRES(MTY,2))DEB=-STRA(MTY,2)+SB*ALOG10(PRF)
      IF(PR.LE.PRES(MTY,4))DEA=-STRA(MTY,1)+SC*ALOG10(PRE)
      IF(PR.GT.PRES(MTY,4))DEA=-STRA(MTY,4)+SD*ALOG10(PRG)
      DEL=DEA-DEB
      IF(NOUT.EQ.0) GOTO 36
      WRITE(6,110) I,DXX,DEL
   36 DELH1=DELH1+DX*DEL
      DXX=DXX+DX
```

Table F-8. Concluded

```
 40 CONTINUE
 50 DELH2=0.0
    IF(NBX.GT.NEL) GOTO 120
    DXX=FLOAT(NBX)*DX-DX/2.
    IF(NOUT.EQ.0) GOTO 65
    WRITE(6,60)
 60 FORMAT(/,1X,'COLLAPSE DISTRIBUTION BELOW FOUNDATION',/,1X,'ELEMENT
   1    DEPTH,FT          DELTA,FT',/)
 65 DO 100 I=NBX,NEL
    MTY=IE(I,1)
    PR=(P(I)+P(I+1))/2.
    PRA=PRES(MTY,2)/PRES(MTY,1)
    PRB=PRES(MTY,3)/PRES(MTY,2)
    PRC=PRES(MTY,4)/PRES(MTY,1)
    PRD=PRES(MTY,5)/PRES(MTY,4)
    PRE=PRES(MTY,1)/PR
    PRF=PRES(MTY,2)/PR
    PRG=PRES(MTY,4)/PR
    SA=(STRA(MTY,2)-STRA(MTY,1))/ALOG10(PRA)
    SB=(STRA(MTY,3)-STRA(MTY,2))/ALOG10(PRB)
    SC=(STRA(MTY,4)-STRA(MTY,1))/ALOG10(PRC)
    SD=(STRA(MTY,5)-STRA(MTY,4))/ALOG10(PRD)
    IF(PR.LE.PRES(MTY,2))DEB=-STRA(MTY,1)+SA*ALOG10(PRE)
    IF(PR.GT.PRES(MTY,2))DEB=-STRA(MTY,2)+SB*ALOG10(PRF)
    IF(PR.LE.PRES(MTY,4))DEA=-STRA(MTY,1)+SC*ALOG10(PRE)
    IF(PR.GT.PRES(MTY,4))DEA=-STRA(MTY,4)+SD*ALOG10(PRG)
    DEL=DEA-DEB
    IF(NOUT.EQ.0) GOTO 80
    WRITE(6,110) I,DXX,DEL
 80 DELH2=DELH2+DX*DEL
    DXX=DXX+DX
100 CONTINUE
110 FORMAT(I5,F13.2,F18.5)
    DEL1=DELH1+DELH2
    WRITE(6,305) DELH1,DELH2,DEL1
305 FORMAT(/,1X,'SOIL COLLAPSE NEXT TO FOUNDATION EXCLUDING COLLAPSE',
   1/,1X,'IN SUBSOIL BENEATH FOUNDATION =',F10.5,' FEET',/,1X,'SUBSOIL
   2 COLLAPSE =',F10.5,' FEET',/,1X,'TOTAL COLLAPSE =',F10.5,' FEET')
120 RETURN
    END
```

C. SETTLEMENT OF COLLAPSIBLE SOIL. The granular soil of the preceding problem was tested in a one-dimensional consolidometer to check for collapse potential. The results of this test were plotted in a compression curve diagram similar to Figure 5-5 and indicated the settlement points shown in Table F-7a, input data for program VDISPL. The output data shown in Table F-7b indicate potential collapse of 0.33 ft or 4 in. beneath the footing.

F-4. Listing

A listing of this program is provided in Table F-8.

APPENDIX G

NOTATION

a_o	Acceleration of vibrations at foundation level, **g**	t_{100}	Time at 100 percent of primary consolidation, days (minutes)
a_{crit}	Critical acceleration, **g**	u_h	Average excess pore water pressure at the bottom of the specimen over the time interval t_1 and t_2, psi
a_{max}	Maximum horizontal acceleration of the ground surface from earthquakes, **g**		
c_u	Uniformity coefficient, D_{60}/D_{10}	u_w	Pore water pressure, tsf
c_v	Coefficient of consolidation, ft²/day	u_{we}	Pore water pressure induced in the soil by foundation loads, tsf
d	Dial reading, in.		
e	Void ratio	u_{wf}	Final pore water pressure, tsf
e_1	Void ratio at beginning of increment	z	Depth, ft
e_c	Void ratio after soaking at $\sigma = 2$ tsf	z_1	Depth of influence of loaded area, ft
e_f	Final void ratio	z_w	Height of column of water above depth z, ft
e_o	Natural or initial void ratio	A_{max}	Maximum displacement of vibration, in.
e_r	Void ratio from which rebound occurs	B	Footing width, ft
e_{LL}	Void ratio at liquid limit	B_p	Diameter of plate, in.
f	Frequency of vibration, rpm	C	Clay content, percent < 2 microns
f_1	Layer thickness correction factor	C_1	Correction accounting for strain relief from embedment, $1 - 0.5\sigma'_{od}/\Delta p \geq 0.5$
f_t	Time-dependent factor relating immediate settlement with settlement at time t		
		C_t	Correction for time-dependent increase in settlement, $1 + 0.2 \cdot \log_{10}(t/0.1)$
f_s	Shape correction factor		
h_a	Average specimen height between time t_2 and t_1, in.	C_α	Coefficient of secondary compression
		CP	Collapse potential
h_e	Equivalent specimen thickness, ft	C_c	Compression index
h_f	Final specimen height, in.	C_r	Recompression index
h_o	Initial specimen height, in.	Cw	Correction for water table depth
k	Constant relating the elastic soil modulus with depth, $E_o + kz$, ksf/ft	Cn	Correction for overburden pressure
		C_N	Depth correction factor for earthquakes
k_d	Coefficient of subsidence	C_s	Swell index
l	Lateral distance between two adjacent points, ft	C_u	Undrained shear strength, tsf
		CPT	Cone penetration test
m	$[(1 + e)/C_c] \cdot \ln 10$	D	Depth of embankment, ft
m'	Shape factor	D_{10}	Grain diameter at which 10 percent of soil weight is finer
q	Vertical pressure applied to soil at bottom of footing, tsf		
		D_{60}	Grain diameter at which 60 percent of soil weight is finer
q_1	Soil pressure from Figure 3-3a using corrected blowcount N' and ratio of embedment depth D to footing width B, tsf		
		D_b	Depth of mat base or stiffening beams below ground surface, ft
q_c	Cone tip resistance, tsf		
q_{oave}	Average pressure in stratum from foundation load, tsf	D_f	Thickness of foundation, ft
		D_r	Relative density, percent
q_p	Plate pressure, tsf	D_w	Depth to groundwater level, ft
r_d	Stress reduction factor for earthquakes	DMT	Dilatometer test
r_m	Scaling factor for earthquakes	E	Elastic modulus, tsf
s	Slope of curve or plot of 1/2 deviator stress versus effective horizontal confining pressure	E_d	Constrained modulus, tsf
		E_D	Dilatometer modulus, tsf
t	Time, days (minutes, years)	E_f	Young's modulus of foundation, tsf
t_{50}	Time at 50 percent of primary consolidation, days (minutes)	E_i	Initial pressuremeter modulus, tsf

E_o	Elastic soil modulus at the ground surface, tsf	$(N_1)_{60}$	Normalized blow count at 60% energy for earthquakes, blows/ft
E_m	Deformation modulus, tsf	N'	Corrected blowcount (Figure 3-3), blows/ft
E_s	Young's soil modulus, tsf (psi)	OCR	Overconsolidation ratio
E_s^*	Equivalent elastic modulus of soil beneath the excavation or foundation, tsf	PI	Plasticity index, percent
		PL	Plastic limit, percent
E_{si}	Elastic modulus of soil layer i, tsf	PMT	Pressuremeter test
E_t	Tangent elastic modulus, tsf	R	Equivalent radius, $\sqrt{LB/\pi}$, ft
E_{ti}	Initial tangent elastic modulus, tsf	R_3	Time-dependent settlement ratio as a proportion of ρ_i during first three years following construction, ≈ 0.3
ER_i	Measured energy ratio for the drill rig and hammer system		
E^*	Theoretical SPT energy applied by a 140-lb hammer falling freely 30 in., 4200 in.-lb	R_t	Time-dependent settlement ratio as a proportion of ρ_i for each log cycle of time after three years, ≈ 0.2
E_i	Available energy, in.-lb	R_{po}	Radius of pressuremeter probe, in.
F_{RD}	Rebound depth factor	R_ϵ	Strain resistance
F_{RS}	Rebound shape factor	R_D	$(1 - v_s)/[(1 + v_s)(1 - 2v_s)]$, relates E_s to E_d
G	Shear modulus, tsf	S	Swell under confinement, percent
G_{eff}	Effective shear modulus at earthquake-induced shear stress, tsf	S_f	Free swell, percent
		S_{max}	Maximum potential heave, ft
G_{max}	Maximum shear modulus, tsf	SPT	Standard penetration test
G_s	Shear soil modulus, tsf	S_{min}	Minimum potential heave, ft
H	Depth of stratum below footing to a rigid base or layer thickness, ft	S_{RE}	Undrained elastic rebound, ft
		T_v	Consolidation time factor
H_e	Equivalent compressible soil height, ft	U_t	Degree of consolidation of the compressible stratum at time t, percent
H_j	Thickness of stratum j, ft		
H_w	Wall height, ft	W	Water content, percent
I	Influence factor for infinitely deep and homogeneous soil in Perloff procedure	W_n	Natural water content, percent
		Z_a	Depth of active zone for heave, ft
I_c	Center influence factor in Perloff procedure, compressibility influence factor	α	Correction factor for subgrade soil in Perloff procedure
I_e	Edge influence factor in Perloff procedure	α_c	Correlation factor depending on soil type and cone bearing resistance
I_w	Influence factor, $\pi/4$ for circular plates		
I_z	Depth influence factor	α_o	Parameter applied in the Alpan method (Figure 3-1a)
I_{zi}	Depth influence factor of soil layer i		
I_p	Peak influence factor	β	Angular distortion
L	Span length or length of footing, ft	β_{max}	Maximum angular distortion
L_{SAG}	Span length with center depression, ft	β_v	Coefficient of vibratory compaction
L_{HOG}	Span length with center heave, ft	γ	Saturated unit weight of soil mass, ton/ft^3
K	Bulk modulus, tsf	γ_c	Cyclic shear strain
K_c	Correlation factor	γ_d	Dry density, lbs/ft^3
K_o	Coefficient of lateral earth pressure at rest	γ_{do}	Initial dry density, lbs/ft^3
K_R	Relative stiffness between soil and foundation	γ_{eff}	Effective cyclic shear strain induced by earthquake
LL	Liquid limit, percent		
M	Magnitude of earthquake	γ_w	Unit weight of water, 0.031 ton/ft^3
N_{ave}	Average blowcount/ft in depth H	γ'	Effective unit wet weight, ton/ft^3
N_{col}	Number of columns in a diagonal line on the foundation	Δ	Deflection, ft
		Δ_a	Allowable differential movement, ft
N_k	Cone factor	Δe	Change in void ratio
N	Average blowcount per ft in the stratum, number of blows of a 140-lb hammer falling 30 in. to drive a standard sampler (1.42" ID, 2.00" OD) 1 ft	Δp	Net applied footing pressure, tsf
		ΔP	Change in pressure measured by pressuremeter, tsf
N_c	Number of cycles	P'_{ave}	Average effective bearing pressure, $q_{oave} + \sigma'_{oave}$, tsf
N_i	Blowcount by Japanese standards, blows/ft		
N_m	Blowcounts measured with available energy E, blows/ft	ΔR_{pm}	Change in radius from R_o at midpoint of pressuremeter curve, in.
N_{60}	Blowcounts corrected to 60% energy, blows/ft		

ΔR_p Change in radius between selected straight portions of the pressure meter curve, in.

Δz_i Depth increment i, ft

$\Delta \sigma$ Increase in effective vertical stress, tsf (psi)

δ Vertical differential movement between two adjacent points, ft

ϵ_c Volumetric strain

$\epsilon_{c,M}$ Volumetric strain for earthquake with magnitude M

ϵ_z Strain in z direction

$\dot{\epsilon}_N$ Rate of strain at number of cycles N_c

$\dot{\epsilon}_1$ Rate of strain at $N_c = 1$

λ Skempton-Bjerrum correction factor

λ_L Lame's constant

λ_d Decay constant

μ_0 Influence factor for depth D in improved Janbu procedure

μ_1 Influence factor for foundation shape in improved Janbu procedure

ν Poisson's ratio

ν_s Soil Poisson's ratio

ρ Total settlement, ft

ρ_c Primary consolidation settlement or center settlement, ft

ρ_{ct} Consolidation settlement at time t, ft

ρ_e Edge settlement, ft

ρ_{col} Settlement of collapsible soil, ft

ρ_e Earthquake settlement, in.

ρ_i Immediate elastic settlement, ft

ρ_{max} Maximum settlement, ft

ρ_s Secondary compression settlement, ft

ρ_t Settlement at time t, ft

ρ_v Vibratory load settlement, ft (in.)

$\rho_{\lambda c}$ Corrected consolidation settlement considering effects of overconsolidation and pore pressure changes from three-dimensional loading, ft

σ_1 Total vertical stress at time t_1, tsf

σ_2 Total vertical stress at time t_2, tsf

σ_a $(\sigma_1 + \sigma_2)/2$, tsf

σ_f Vertical confining pressure, tsf

σ'_{izp} Effective overburden pressure at the depth of I_{zp}, tsf

σ'_{oave} Average effective overburden pressure in stratum H, tsf

σ'_{od} Effective overburden pressure at depth D or bottom of footing, tsf

σ_{oz} Total overburden pressure at depth z, tsf

σ_p Total maximum past pressure, tsf

σ'_{qp} Apparent preconsolidation stress, tsf

σ_{rd} Repeated deviator stress, tsf

σ_{st} Vertical pressure from foundation loads transmitted to a saturated compressible soil mass, tsf

σ_s Swell pressure, tsf

σ'_f Final effective pressure, tsf

σ'_{hz} Effective horizontal pressure at rest at depth z, tsf

σ'_o Effective overburden pressure, tsf

σ'_{od} Effective overburden pressure at bottom of footing, tsf

σ'_m Mean effective pressure, tsf

σ'_P Preconsolidation stress or maximum past effective stress, tsf

τ_{av} Average cyclic shear stress induced by earthquake shaking, tsf

ω Fraction tilt

ω_o Angular rotation, radians/sec

INDEX

134